An Arduino Workshop

From the Nuts and Volts Magazine
Smiley's Workshop Series

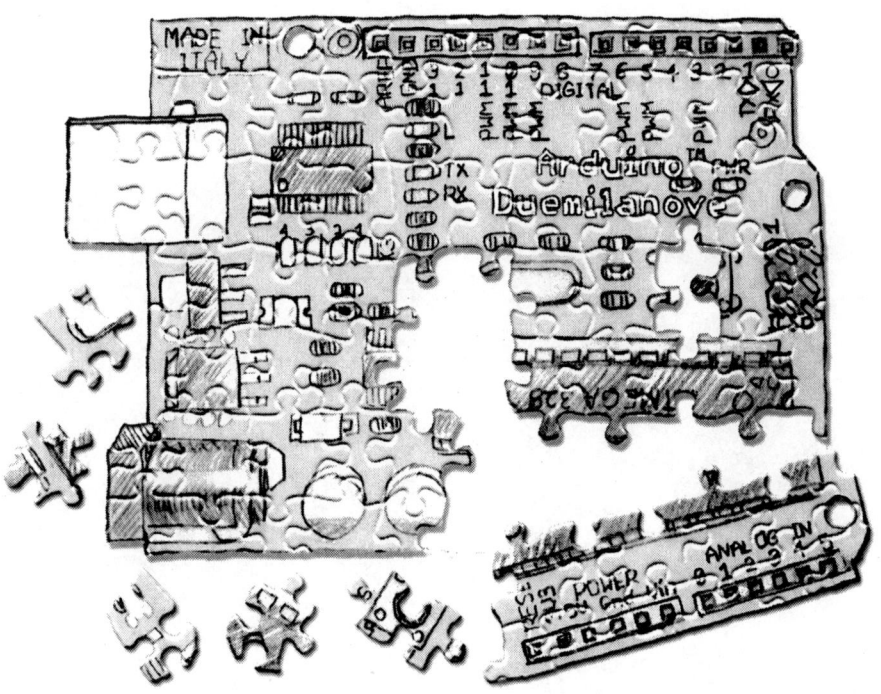

Joe Pardue - SmileyMicros.com

Copyright © 2010 by Joe Pardue, All rights reserved.
Published by Smiley Micros

Smiley Micros
5601 Timbercrest Trail
Knoxville, TN 37909
Email: book@SmileyMicros.com
Web: http://www.SmileyMicros.com

ISBN 978-0-9766822-2-6

Products and services named in this book are trademarks or registered trademarks of their respective companies. In all instances where Smiley Micros is aware of a trademark claim, the product name appears in initial capital letters, in all capital letters, or in accordance with the vendor's capitalization preferences. Readers should contact the appropriate companies for complete information on trademarks and trademark registrations. All trademarks and registered trademarks in this book are the property of their respective holders.

No part of this book, except the programs and program listings, may be reproduced in any form, or stored in a database of retrieval system, or transmitted or distributed in any form, by any means, electronic, mechanical photocopying, recording, or otherwise, without the prior written permission of Smiley Micros or the author. The programs and program listings, or any portion of these, may be stored and executed in a computer system and may be incorporated into computer programs developed by the reader.

NONE OF THE HARDWARE USED OR MENTIONED IN THIS BOOK IS GUARANTEED OR WARRANTED IN ANY WAY BY THE AUTHOR. THE MANUFACTURERS OR THE VENDORS THAT SHIPPED TO YOU MAY PROVIDE SOME COVERAGE, BUT THAT IS BETWEEN YOU AND THEM. NEITHER THE AUTHOR NOR SMILEY MICROS CAN PROVIDE ANY ASSISTANCE OR COMPENSATION RESULTING FROM PROBLEMS WITH THE HARDWARE.

The information, computer programs, schematic diagrams, documentation, and other material in this book are provided "as is," without warranty of any kind, expressed or implied, including without limitation any warranty concerning the accuracy, adequacy or completeness of the material or the results obtained from the material or implied warranties. Including, but not limited to, the implied warranties of merchantability and fitness for a particular purpose are disclaimed. *Neither the publisher nor the author shall be responsible for any claims attributable to errors, omissions, or other inaccuracies in the material in this book. In no event shall the publisher or author be liable for direct, indirect, special, exemplar, incidental, or consequential damages in connection with, or arising out of, the construction, performance, or other use of the material contained herein. Including, but not limited to, procurement of substitute goods or services; loss of use, data, or profits; or business interruption however caused and on any theory of liability, whether in contract, strict liability, or tort (including negligence or otherwise) arising in any ay out of use, even if advised of the possibility of such damage. **In no case shall liability be implied for blindness or sexual impotence resulting from reading this statement although the author suggests that if you did read all this then you really need to get a life.***

Table of Contents:

Chapter 1: Introduction 13
   So what is Arduino? 14
   Smiley Micros Arduino Projects Kit 15
      A Few Definitions. 16
   Genesis of Arduino 18
      Processing 18
      Wiring 18
      Arduino 19
      Smiley's Workshop Nuts&Volts Articles 19
      USB Serial Port 19
   How is this book organized? 20
      Prerequisites: 20
   IMHO (In My Humble Opinion) is implied 21
      Errors happen 21
      "We really value your business, but all our lines are busy …" 21
   Is Arduino the end or the beginning? 21
   Where to get your Arduino 22
Chapter 2: Arduino Quick Start Guide 23
      The Arduino Integrated Developers Environment 23
      Select the Arduino Duemilanove with ATmega328 26
      Load the Blink Program, uhh… Sketch 27
         Blink Source Code: 28
      Verify the Sketch (Compile the Program). 28
      Uploading the sketch (program) to the Arduino board 30
   Getting Help 30
      Using the Arduino Language Reference 31
      Using Internet Forums 33
         Yes, you will get flamed 33
Chapter 3: Playing with software – Part 1 35
   How do we write a program? 35
   How an Arduino Program is structured 36
      Structure of the Arduino Blink example 37
   Learning the programming language rules 38
   An Arduino Light Switch 39
      What do we want our program to do 39

- Set up the hardware 40
  - Light_Switch - source code: 41
- Some of the rules 43
  - Comments 43
  - Functions 43
  - Expressions, Statements, and Blocks 45
  - Flow Control 45

Chapter 4: Hardware Prototyping 47
- Introduction to Breadboards 47
  - How a breadboard works 47
- An Introduction to Schematics 50
  - Schematic Symbols 51
- Using a breadboard with the Arduino 52
- The Arduino Learning Platform (ALP) 53
  - ALP Base 55
  - Arduino Learning Platform Storage Box 56

Chapter 5: Some Simple Projects 59
- Digital Input and Output 59
  - DIP Switch and LEDs 59
    - The DIP Switch 60
    - DIP to LED Source Code: 61
  - Cylon Eyes the Arduino Way 62
    - Cylon Eyes 1 Source Code: 62
  - Output sound – piezo music 66
  - Sounds Components, Schematic, Layout 67
  - Tunes 68
    - Happy Birthday Source Code: 70
- Analog 72
  - Using PWM to Fade an LED 72

Chapter 6: Playing with electricity 75
- Electricity is dangerous (well, duh!) 75
- Electric Measurements 76
- Electric Potential Difference = Voltage 77
- Electric Current = Amps 78
- Electric Resistance = Ohms (Ω) 79
- Ohm's Law 79
- Circuits 80

Short Circuits 81
Electrifying Experiments 82
    An Arduino Volt Meter 82
        Installing the Arduino Volt Meter on your PC 83
        Opening the Arduino in the Volt Meter 83
        AVM_Test Source Code: 84
    Voltage across resistance 84
    Variable Resistance: the Potentiometer 88
    LED Dimmer 93

Chapter 7: Playing with software – Part 2 97
Microcontroller I/O Ports 97
    DIPLED_With_Ports Source Code: 99
Redo Cylon Eyes using ports 100
    Cylon Eyes 2 Source Code: 100
Operators 102
Some equals are more equal than others 104
There are exactly 10 types of people in the world. Those who understand binary and those who don't. 104
    Bits 104
    Bytes 105
Bitwise Operators 107
    The TRUTH about numbers 107
ORing 108
ANDing 109
Setting and Clearing Bits 110
XORing 111
NOTing 111
Bitwise versus Boolean Operators 112
    Launch_Control Source Code: 114
Shift Operators 115
Masks and Macros: Using Named Bits 116
    CylonOptometery Source Code: 119

Chapter 8: Communicating with a PC 129
ASCII and the char data type 129
    So what's the big deal about ASCII anyhow? 130
Sending an receiving numbers 130
    Number_Commander Source Code: 132

Arduino Jukebox Source Code:	134
Getting real with serial input	139
ASCII_To_Integer Source Code:	141
Command Demonstration	143
Command_Demo Source Code:	144
Chapter 9: Sensors	**149**
Light Sensor	150
Light Sensor Components, Schematic, Layout	151
A Word or Two about Storing and Showing Sensor Data	152
A quick introduction to signed decimal numbers	152
Showing integer data as signed decimal fractions	153
Converting Centigrade to Fahrenheit	154
The LM35 Temperature Sensor	155
Temperature Sensor Components, Schematic, Layout	157
LM35_Temperature Source Code:	159
Infrared Object Detection	162
But is it Light?	162
Some Common Uses for IR Sensors	164
IR Reflective Object Sensor – the QRD1114	165
Making IR visible	165
IR doesn't know black from white	166
Show invisible IR intensity with visible red LED	166
Tomato Soup Can Counter	170
Tomato Soup Can Counter Software	171
ReadPot Source Code:	171
Tomato_Soup_Can_Counter Source Code:	173
Chapter 10: Simple Motor Speed Control	**177**
Using external interrupts to detect edges	178
Edge Detect Interrupt Software	180
Edge_Detect_Interrupt Source Code:	180
Optical Isolation of voltage	182
Optical Isolation Component, Schematic, Layout	184
Optical Isolation Source Code	185
PWM_Test Source Code:	185
How our DC motor works	188
Diode to suppress voltage spikes	191
Powering the motor	192

  Using PWM to control the motor speed 193
  Building the breadboard circuit 194
  Using an encoder wheel to measure the motor speed 197
  Simple motor speed control with digital feedback 199
    Simple_Motor_Speed_Control Source Code: 199
  Now what? 203
Appendix 1: ASCII Table 205
Appendix 2: Decimal, Hexadecimal, and Binary 207
Index 209

Table of Figures:

Figure 1: The Arduino Duemilanove 13
Figure 2: The Arduino Projects Kit 15
Figure 3: Arduino Projects Kit Parts List 16
Figure 4: Bogus Security Warning 23
Figure 5: cmd.exe window 24
Figure 6: Arduino desktop icon 24
Figure 7: The Arduino Integrated Developers Environment 25
Figure 8: Select the Duemilanove board 26
Figure 9: Selecting the Blink example 27
Figure 10: Verify the sketch 29
Figure 11: Compile 29
Figure 12: Done compiling 29
Figure 13: Upload to the I/O board 30
Figure 14: Help\Reference 31
Figure 15: Arduino Language Reference 32
Figure 16: pinMode function reference 33
Figure 17: Schematic – Pushbutton switch and LED 39
Figure 18: Drawing - pushbutton switch and LED 40
Figure 19: Breadboard front and back 48
Figure 20: Back with 5-position vertical and 50-position clips 48
Figure 21: A 5-position clip 49
Figure 22: Breadboard cross-section with LED, resistor, and wire 49
Figure 23: Layout Drawing and EAGLE Schematic for LED 51
Figure 24: Arduino pin-out schematic 52
Figure 25: The ALP base. 53
Figure 26: Slide it in the ALP box 54
Figure 27: Scruffy protective box 54
Figure 28: Construction site 55
Figure 29: Stick on Velcro squares 56
Figure 30: Start assembly 56
Figure 31: Glue the sides together as shown 57
Figure 32: Front before taping lid 57
Figure 33: DIP switch and LEDs drawing 59
Figure 34: 8-position DIP switch. 60
Figure 35: DIP switch and LED schematic 61

Figure 36: Piezo element layout..66
Figure 37: Piezo element illustration 67
Figure 38: Piezo element schematic..68
Figure 39: Note table...68
Figure 40: 'c' note waveform..69
Figure 41: PWM Fade 73
Figure 42: Learning the shocking truth about electricity. 75
Figure 43: Bucket voltage metaphor 77
Figure 44: Amp metaphor 78
Figure 45: Flow of conventional current..80
Figure 46: Arduino Volt Meter 82
Figure 47: Select your Arduino...83
Figure 48: Schematic of resistors in series..85
Figure 49: Layout of resistors in series 86
Figure 50: AVM measuring resistor divider 87
Figure 51: Potentiometer...88
Figure 52: Potentiometer metaphor and real circuit..89
Figure 53: The potentiometer wiper..91
Figure 54: Potentiometer to Arduino schematic 92
Figure 55 Arduino reads potentiometer 92
Figure 56: Pot, LED, and Arduino schematic 93
Figure 57: Pot, LED, and Arduino drawing 94
Figure 58: AVM AT 1.62..95
Figure 59: Mapping ATmega168/328 to Arduino pins...98
Figure 60: Arduino Pin9 is ATmega328 PortB Pin1 99
Figure 61: DIP Switch Use 117
Figure 62: Select the Arduino IDE Serial Monitor 133
Figure 63: Number Commander in the Serial Monitor...134
Figure 64: Arduiono IDE Serial Monitor ASCII_To_Integer.....................................143
Figure 65: Command Demo..144
Figure 66: CdS light sensor layout..149
Figure 67: CdS light sensor...150
Figure 68: CdS light sensor schematic..151
Figure 69: Centigrade mapped to Fahrenheit..154
Figure 70: Temperature, voltage, and ADC ranges 156
Figure 71: LM35 temperature sensor..157
Figure 72: Temperature sensor schematic...158

Figure 73: Temperature sensor layout drawing 159
Figure 74: Output in Arduino Serial Monitor 161
Figure 75: Cute L'il Bunny has her last thought. 162
Figure 76: Dog IR thermoscan 163
Figure 77: Digital camera 'sees' TV remote IR 164
Figure 78: QRD1114 Cross-section 165
Figure 79: QRD1114 Reflective Object Sensor 166
Figure 80: Bending the QRD1114 legs 167
Figure 81: Object Detector Schematic 168
Figure 82: IR Object Detector Layout 168
Figure 83: IR Object Detector Photo 169
Figure 84: Tomato soup can counter prototype 170
Figure 85: ADC readings of finger movements 172
Figure 86: Can count on Arduino serial monitor. 175
Figure 87: Simple Motor Speed Control 177
Figure 88: Edge detection schematic 180
Figure 89: Edge Detect Interrupt counter 181
Figure 90: 4N25 Optically Coupled Isolator 182
Figure 91: QRD1114 and 4N25 schematic symbols 183
Figure 92: Optoisolation Test Layout 184
Figure 93: Optoisolator Test Circuit 185
Figure 94: DC Motor Principles 188
Figure 95: DC motor dismantled 190
Figure 96: Diode 191
Figure 97: PWMs on oscilloscope 193
Figure 98: DC Motor 195
Figure 99: Power Transistor 196
Figure 100: Motor Speed Control Schematic 196
Figure 101: Motor Speed Control Layout 197
Figure 102: Encoder Wheel 198
Figure 103: Program Serial I/O 199

Chapter 1: Introduction

Chapter 1: Introduction

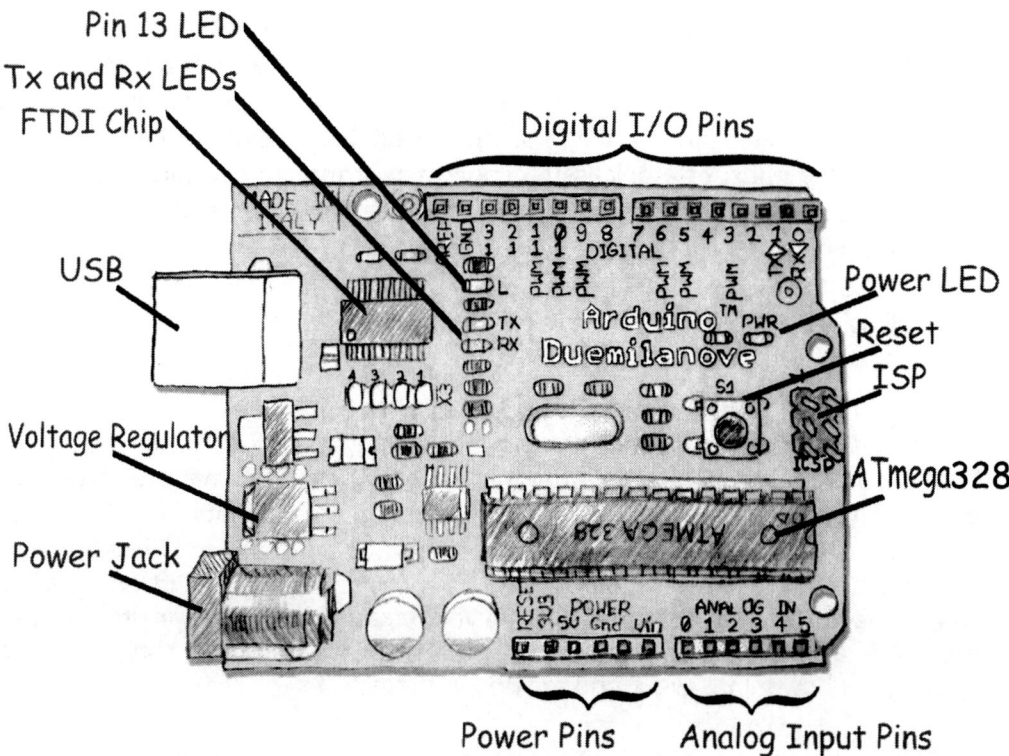

Figure 1: The Arduino Duemilanove

Work? I put 'workshop' in the title so that you would know that you are going to get your hands dirty. This is not a book to feed ideas into your brain while you relax on a deck chair drinking Piña Coladas. It is a book where you spread your Smiley Micros Arduino Projects Kit out on a table next to a PC and build stuff to help you learn how to **play** with microcontrollers and electrical hardware.

Play? Yes it is not just work, and the real emphasis here is play. If it ain't fun, why do it?

Chapter 1: Introduction

One of the really cool things about the Arduino is that you can start playing with it without understanding much of what you are doing. We will begin with this concept in mind and play with the Arduino and some of the stuff in the Projects Kit. Later, we will begin to learn about the details of how the software and hardware work. Be aware, however, that the presentation of topics may not seem entirely logical, because, well… it isn't particularly logical. I organized this book thinking about how I might present various topics to a novice in a series of hands-on workshops. This means that you may first learn about a certain software operator in the midst of a discussion about measuring temperature. You will find some of what is presented is intuitively obvious and other stuff may cause you WTF moments.

So what is Arduino?

Massimo Banzi begins his book, *Getting Started with Arduino:*
"A few years ago I was given the very interesting challenge: teach designers the bare minimum in electronics so that they could build interactive prototypes of the objects they were designing." He summarizes his philosophy of 'learning by tinkering' with a quote from www.exploratorium.edu/tinkering: *"Tinkering is what happens when you try something you don't quite know how to do, guided by whim, imagination, and curiosity. When you tinker, there are no instructions – but there are also no failures, no right or wrong ways of doing things. It's about figuring out how things work and reworking them."*

Arduino provides a great toolset for designers, tinkers, and even some of us surly old engineers who sometimes just want to play with an idea. The genius of Arduino is that it provides just enough access to get specific tasks done without exposing the underlying complexities that can be truly daunting for folks new at this stuff.

Chapter 1: Introduction

Smiley Micros Arduino Projects Kit

You can get the Arduino Projects Kit from www.smileymicros.com.

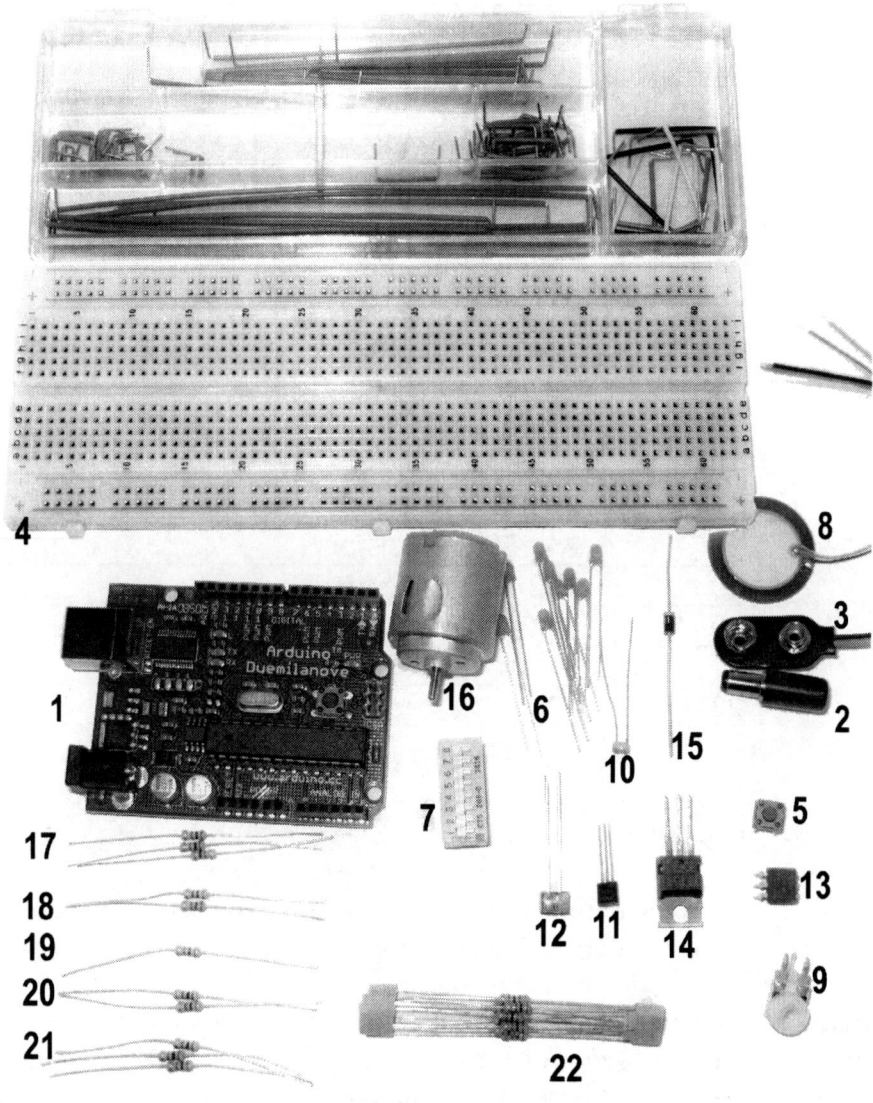

Figure 2: The Arduino Projects Kit

Chapter 1: Introduction

Part #	Arduino Projects Kit
1	Arduino Duemilanove (ATmega328)
2	DC Plug 2.1mm(ID)5.5(OD) 9.5mm
3	9V Battery Snap
4	Breadboard + Jumper wire kit
5	Push button
6	10 - LED Red 3mm
7	DIP Switch SPST 8pos
8	Piezo element
9	Potentiometer 10K
10	CdS Light Sensor
11	LM35 Temperature Sensor
12	IR Reflector QRD1114
13	Optoisolator 4n28
14	Power Transistor TIP115
15	Diode 1N4001
16	Motor DC 6-12V
17	3 -100 Ω Resistor
18	2 - 150 Ω Resistor
19	1- 270 Ω Resistor
20	1 - 2.2k Ω Resistor
21	3- 10k Ω Resistor
22	10 - 1k Ω Resistor

Figure 3: Arduino Projects Kit Parts List

A Few Definitions.

Sensor – a computer-controlled mechanism that measures or detects events in the real world such as motion, light, or voltage. For example: an IR optical sensor used to detect the motion of a computer mouse.

Actuator – a computer controlled mechanism that causes a device to be turned on or off, adjusted or moved. For example: the devices that spin a DVD, move the sensing head, and turn the read/write laser on/off.

Micro (microcontroller) – a single integrated circuit that contains a computer, memory, and peripherals used with sensors and actuators to provide **control** of

Chapter 1: Introduction

real-world events. For example: an Atmel AVR microcontroller such as the ATmega328 can be used to sense button presses, motor rotation, and temperature and use that input to control the voltage to the cavity magnetron (makes the microwaves) in a microwave oven.
Real World – for our purposes, this is everything outside our micro and associated sensors and actuators, though to be completely honest, I gave up on trying to define 'real world' years ago.
Physical Computing (Embedded Systems) – a combination of micro(s), sensor(s), and actuator(s) designed for some specific control function and 'embedded' into a specific device, usually requiring little human input. An example would be the air/fuel mixture control system in an automobile engine. This term is often used in contrast to a 'general-purpose computer' such the PC or Mac, which can do a lot of different things and are designed for intense human interaction.
Sketch (program) – a sequence of human readable text statements following the rules of a programming language that can be converted and uploaded to a micro providing instructions as to how it will make decisions about using its sensors and actuators.
Verify (compile) – the process of converting a sketch to instructions suitable for the micro. The process provides feedback if the conversion cannot be completed due to errors.

While using Arduino is relatively simple, it is built on top of some things that can quickly become complex if you dig a little below the surface.

Arduino Hardware Resources (see Figure 1)
- 14 Digital I/O pins, six of which can be configured as analog output (PWM) pins
- 6 Analog Input pins
- Pins are on female sockets that are easy to wire to a breadboard
- USB serial port with +5V bus power through a PTC fuse.
- Serial transmission and reception LEDs.
- May be switched to regulated external power socket.
- ATmega328 with a pre-programmed bootloader.
- ISP (In-System Programming) header
- Reset button

Chapter 1: Introduction

Genesis of Arduino

Verily in the olden days, artists wailed for lack of entry to the new land. And they were heard by great prophets who showed them a path: Processing. And lo, it came to pass that Processing begat Wiring which begat Arduino that, the time being ripe, begat a whole nation of xxx-duino derivatives. This dear brethren is the story:

Processing

You have probably seen some of the cool (or annoying depending on your viewpoint) graphic Java applets that bless (or infest) Internet browsers. But things like those 'too darn cute' rotating bouncing cubes with different pictures or messages on each face did show that you could use Java to build some very interesting graphic things, and artists took notice. The only problem is that artists tend not to be programmers and vice versa. Casey Reas and Benjamin Fry while in the Aesthetics and Computation Group at the MIT Media Lab took note of the potential for Java to be used by artists to do artsy stuff and decided to create a simpler interface that would be more artist friendly. They developed Processing: a programming language, development environment, and online community [www.processing.org]. The syntax looks a lot like C (but then so does Java if you squint). The Arduino Integrated Development Environment (IDE) is derived from the Processing IDE.

Remember that Processing was designed for artists. It substitutes 'sketch' for 'program' and 'verify' for 'compile'. If you are at all interested in art, be sure and go to their web site and click on the exhibition tab.

Wiring

Wiring was started by Hernando Barragán at the Interaction Design Institute Ivrea in Italy and builds on Processing to provide an 'electronics prototyping input/output board for exploring arts and tangible media.' The Arduino folks smash this mouthful into 'physical computing' and I'm temped to redub it 'playing with electrons' because it makes things so much easier than many earlier systems that the concept of 'play' is not out of the question. The Wiring board is based on the ATmega128 and the language has libraries built around that device's resources and tasks you might like to do with them (like blinking LEDs or monitoring a microphone). The good news is that this is a feature rich platform

Chapter 1: Introduction

and the bad news is that all those riches cost money. So enter the Arduino designed to be affordable for students with the cost goal being the same as going out and getting a pizza.

Arduino

Arduino is built directly out of Wiring except that it uses a more affordable board, and has a very active online community. Massimo Banzi (www.tinker.it) thought up the name Arduino and, if the story is true, it was christened five minutes before the first board was to go into production. Massimo was on the phone with the PCB fabricator who wanted a name for the board and Massimo suggested the name of a bar that some of the developers once frequented. And while the name may be somewhat accidental, it is a masterstroke of Google-ability especially compared to the names Processing or Wiring (try Goolge-ing the three and see what I mean). Since this is all open-source, it didn't take long for lots of folks to make their own versions of the board and use the –'duino' in the name as if it was a valid suffix. This apparently irritates some Italians who note, for instance, that the board name Sanguino would be what and Italian would shout if he was bleeding. There was a bit of debate on the Arduino forum about PCB designs perhaps straying too far from the original concept and making it difficult to provide support so several of the originators copyrighted the name and allow its use for boards that pass muster. Naturally a conspiracy theory rebellion occurred and some folks mounted their high horses and came up with the name Freeduino as a 'free' alternative (www.freeduino.org).

Smiley's Workshop Nuts&Volts Articles

This book is based very loosely on a series of articles in Nuts&Volts magazine: Smiley's Workshop parts 9 through 17. Those articles have been reorganized, rewritten, and expanded for this book so expect some overlap if you read them, but be sure and note that those articles aren't a substitute for this book.

USB Serial Port

By the way, the USB serial port on the Arduino uses the FTDI FT232R chip which was discussed in detail in the article "The Serial Port is Dead, Long Live the Serial Port' by yours truly in the June 2008 issue of Nuts&Volts (you can get a copy on www.smileymicros.com). And if you come to really like that chip you might want to get the book *Virtual Serial Programming Cookbook*" (also by yours truly) and associated projects kit from the Smiley Micros website.

Chapter 1: Introduction

How is this book organized?

I find it very difficult to organize information to help folks understand microcontroller hardware and software in a reasonably logical sequence for both hardware and software at the same time. It seems that I have to choose either hardware or software as the central theme and then let the other be a supporting mishmash. In my *C Programming for Microcontrollers* book, I chose the programming language as the guide and the hardware got slapped in, to support software examples. In this book the hardware is the guide and the software is crammed in as needed.

Prerequisites:
1. You have access to a PC with a USB port and a high-speed Internet connection. I personally think that a good Internet connection is the best learning tool ever invented, so get one.
2. You know simple math concepts. If you are given the equation $X = Y/2$ and that Y is equal to 10 and you can tell me what X is, then you should do just fine.
3. Don't even start unless:
    a. You have near saint-like patience. Even though this stuff is as easy as I can make it, and loads of fun for most folks, it is still hard and will take lots of time and work.
    b. You love puzzles. You will rarely get one of the projects working first time without having to puzzle out something.

I include the #3 prerequisite since I am a frequent and somewhat surly contributor to the forum on AVRFreaks and several times a week I read a thread posted by someone whose problem IMHO (see next page) is that the person lacks patience or doesn't like solving puzzles. Such folks really should be doing something else with their time. I love this stuff, and I have to assume that if you are reading this, then you are one of the folks who think you might love it also. But if your first instinct on encountering a problem is to go on an Internet forum and plead for help – do yourself a favor and don't even start.

Chapter 1: Introduction

IMHO (In My Humble Opinion) is implied

I get a lot of really nice compliments for my writing but occasionally someone seems to think that I ought to be burned at the stake for some heresy I dared speak. I have fun writing and occasionally I'll be flip with an issue that turns out to be someone else's sacred cow. When I realize that I may be moving into a sensitive topic, I try to remember to insert IMHO to remind folks that I haven't put myself forward as some kind of Guru-Swami-Know-It-All, but just an ordinary smuck having fun and trying to share the fun with others as best I can.

Errors happen

Also I make mistakes (lots of mistakes), and as much as I'd like to say I'm sorry, I won't because I'd be lying. The only people who don't make mistakes are the ones who don't do anything. I suggest that you look on www.smileymicros.com at the web page for this book and review the Errata before reading further. If you find an error that is not in the Errata, then please notify me at: error@smileymicros.com.

"We really value your business, but all our lines are busy …"

As much as I'd like to answer questions from everybody who reads this book, I simply don't have the time. Probably more import is that I think of this work as being part of a community effort. If you have a question, then it is likely that other folks will have the same question, so it is far better that the question gets asked and answered in a public forum such as the ones on www.arduino.cc or www.avrfreaks.net so that everyone has access to the answer. If I see that a question gets asked repeatedly, I'll post a FAQ on the Smiley Micros web page associated with this book. Be sure and mention the book in the thread title to increase the chances I'll see it.

Is Arduino the end or the beginning?

One thing to note is that the book goes further than many folks may consider standard for an Arduino workshop, and by the end we will be moving well beyond the Arduino simplifications and into topics that I think will help you advance in your study of how to use microcontrollers in a more general sense.

Chapter 1: Introduction

I suspect most folks who might describe themselves as artists and mainly want to do things like blink LEDs, read switches and spin motors will be more than happy to put up with Chapters 1 to 6, and then start skimming the text and copying projects, since it starts getting more detailed from Chapter 7 onward. I also suspect that some folks who are thinking about more advanced work, like going into a University level program in Mechatronics or EE will want to go all the way to the end of the book to get a leg up on the work they will be doing later.

The Arduino fills the stated goal by enabling designers to build interactive prototypes with the bare minimum of electronic knowledge - and actually exceeds it in many ways. And while it is an admirable goal, it is not my goal. My goal is to help folks get started down this fascinating path and to me the Arduino is a good start, but by no means an end point unto itself.

This isn't a linear prescribed program leading to any kind of certification, grade, or degree so you are the one who has to decide what is enough and what is too much.

Where to get your Arduino

People who fret about the cost of buying the original Arduino from the folks who designed it versus saving a few bucks buying a knock-off should do some soul searching over their priorities. I have no financial arrangements with the manufacturers of the original Arduino (the one with the picture of Italy on it) but I use it in my Arduino Projects Kit and recommend it as one way to give back to the community that created it in the first place.

Chapter 2: Arduino Quick Start Guide

The Arduino Integrated Developers Environment

You can download the Arduino software from www.arduino.cc. This book is based on release 0015 working on a Windows XP PC (also tested with Vista Home Basic).

The Arduino folks seem to be posting new releases several times a year, so there is a possibility that the latest version available will have some changes that are not compatible with the release used in this book. If that turns out to be the case you can get release 0015 from my website.

After downloading to your directory, you can go ahead and click on ..\arduino-0015\Arduino.exe and see what happens; it might run on your system. But in order to run the Processing IDE on my Windows XP machine I have to click on the ..\arduino-0015\run.bat file. Windows doesn't like that so it displays a security warning (Figure 4: Bogus Security Warning). But to heck with Windows and its so-called security (like they would know security), just click 'Run'.

Figure 4: Bogus Security Warning

Chapter 2: Arduino Quick Start Guide

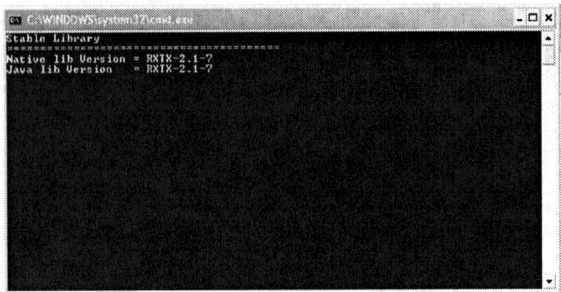

Figure 5: cmd.exe window

Which will open a 'cmd' window (Figure 5: cmd.exe window) and in a little while Windows will run Java for you and display the Processing IDE configured for the Arduino (Figure 7: The Arduino Integrated Developers Environment).

In Vista Home Basic, you only need to click on the Arduino icon (Figure 6: Arduino desktop icon). Don't be alarmed when your screen goes black for a moment, this is Java doing something weird. It will happen again when you close Arduino.

Figure 6: Arduino desktop icon

If you system doesn't fit any of these instructions, just go to www.arduino.cc and download the latest and greatest version and follow their instructions.

24

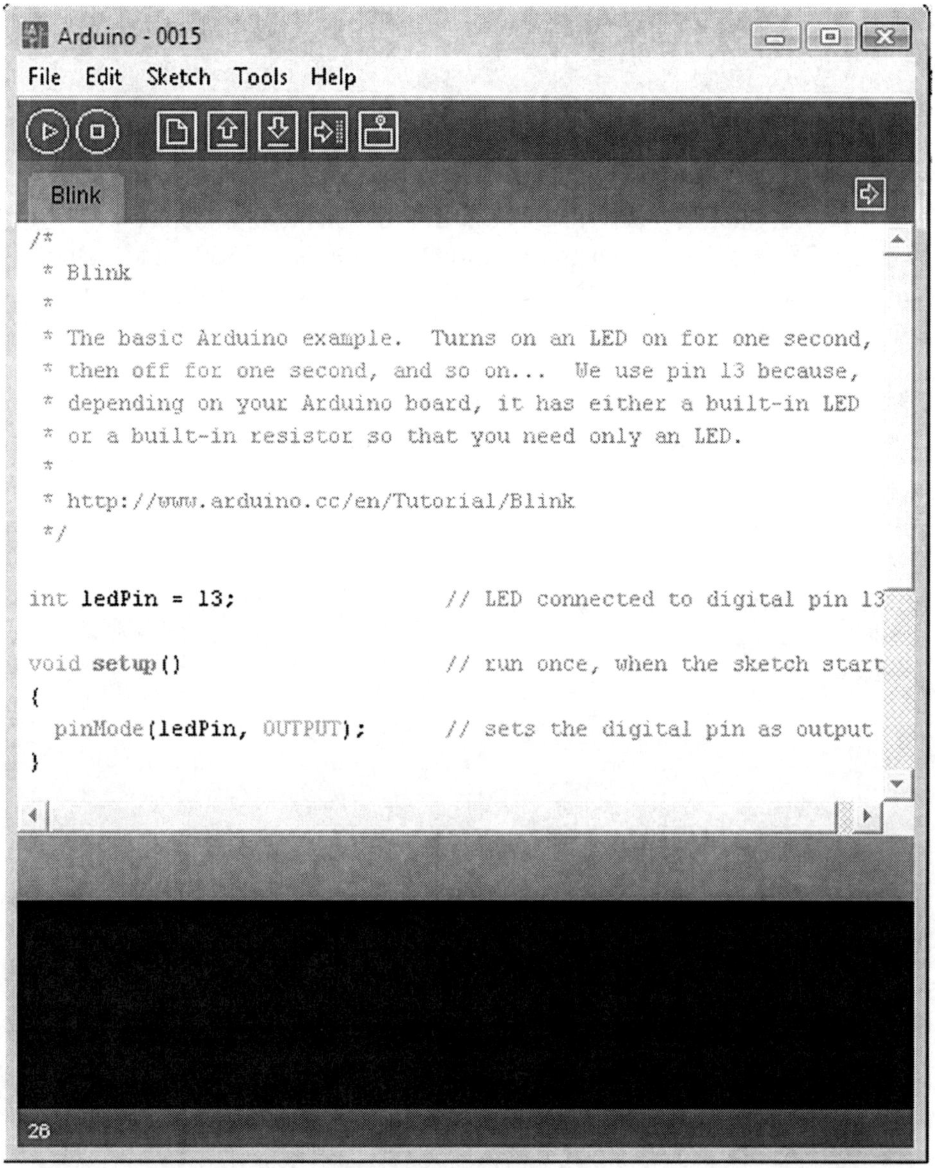

Chapter 2: Arduino Quick Start Guide

Figure 7: The Arduino Integrated Developers Environment

Chapter 2: Arduino Quick Start Guide

Select the Arduino Duemilanove with ATmega328

Our Arduino Projects Kits have caused a little confusion because it uses the Duemilanove (Italian for 2009) with the ATmega328 (double the memory at the same price) instead of the older ATmega168. These processors have 32 or 16-kilo bytes of memory respectively so they require a different setting in the Arduino IDE. For our board, open the Tools/Boards menu and select the Arduino w/ ATmega328. Also note that the bootloader* runs at 57600 baud, which is faster than the older Arduino bootloader and seems to be confusing some folks on the Arduino forum, so be careful.

Open the 'Tools' menu item, select 'Board', and then 'Arduino Duemilanove w/ATmega328' as shown in Figure 8: Select the Duemilanove board.

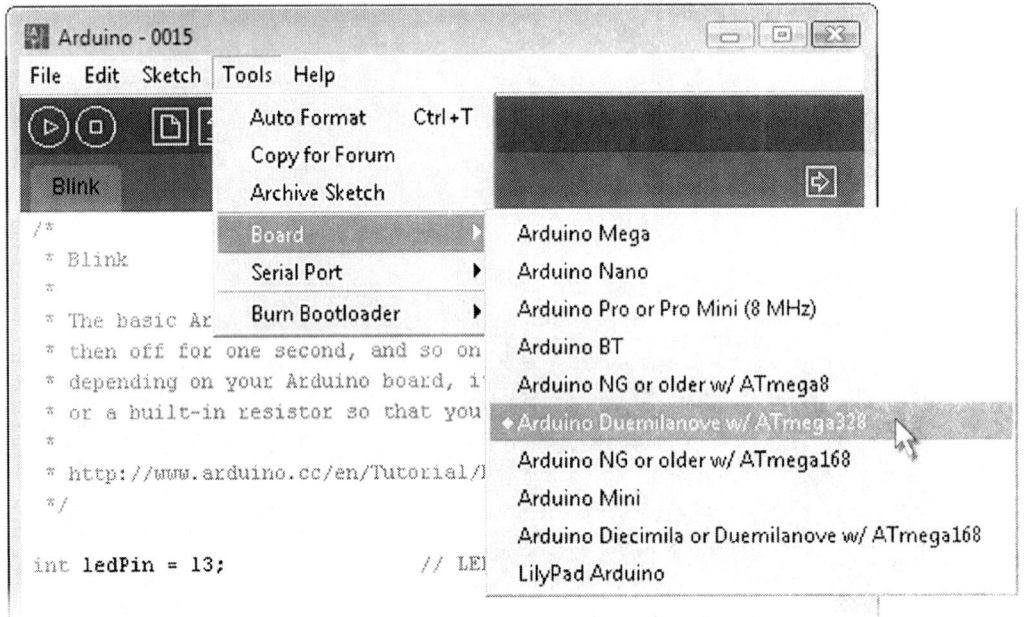

Figure 8: Select the Duemilanove board

*A bootloader is software on the microcontroller that is used to upload application programs from the PC to the microcontroller.

26

Chapter 2: Arduino Quick Start Guide

Load the Blink Program, uhh… Sketch

The Arduino Duemilanove has a built in LED with a resistor attached to I/O pin13 (see Figure 1: The Arduino Duemilanove), so we will start with a program (okay, sketch!) to blink that LED.

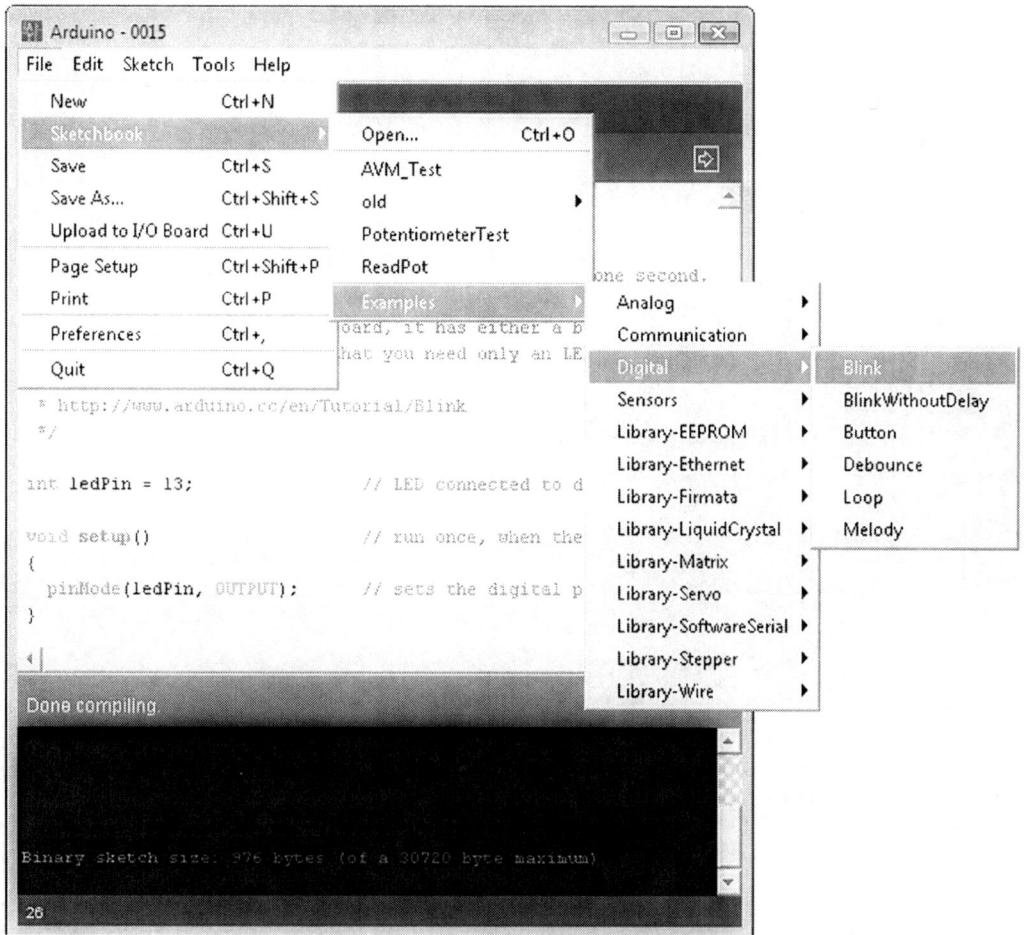

Figure 9: Selecting the Blink example

The following sketch was copied from the Arduino-0015 menu: File/Sketchbook/Examples/Digital/Blink (Figure 9: Selecting the Blink example) and modified to better fit in the book.

27

Chapter 2: Arduino Quick Start Guide

Blink Source Code:
```
/*
 * Blink
 *
 * The basic Arduino example.
 * Turns on an LED on for one second,
 * then off for one second, and so on...
 * We use pin 13 because, depending on your
 * Arduino board, it has either a built-in
 * LED or a built-in resistor so that you
 * need only an LED.
 *
 * http://www.arduino.cc/en/Tutorial/Blink
 */
// LED connected to digital pin 13
int ledPin = 13;

// run once, when the sketch starts
void setup()
{
  // sets the digital pin as output
  pinMode(ledPin, OUTPUT);
}

// run over and over again
void loop()
{
  // sets the LED on
  digitalWrite(ledPin, HIGH);
  delay(1000); // waits for a second

  // sets the LED off
  digitalWrite(ledPin, LOW);
  delay(1000); // waits for a second
}
```

And, no, you aren't expected at this point to understand any of the software. We will discuss those details later, but let's first run the code and blink that LED.

Verify the Sketch (Compile the Program).

The Arduino folks also decided to call compile 'verify'. Instead of 'compiling the program' we are verifying the sketch'. [Artists... go figure] When you click on the button shown in Figure 10: Verify the sketch, the IDE uses the WinAVR

Chapter 2: Arduino Quick Start Guide

toolset for the GCC C compiler, and if everything goes okay – meaning it compiles without error, you don't have to know what WinAVR is, how it works, or even that it is there. We'll look at the situation involving errors later.

Figure 10: Verify the sketch

After you click verify you will notice some activity at the bottom of the Processing window. Figure 11: Compile shows the bottom block if the IDE you will see 'Compiling'.

Figure 11: Compile

Figure 12: Done compiling

Chapter 2: Arduino Quick Start Guide

And please don't ask me why they call it 'verify' at the top of the IDE, but 'compile' at the bottom. Hopefully you got the message: "Binary sketch size: 976 bytes (of a 30720 byte maximum" meaning everything is copasetic. If you wonder why there are only a maximum of 30720 bytes available when the ATmega328 has 32768 bytes of memory, it is because 2048 bytes are set aside for the bootloader.]

Uploading the sketch (program) to the Arduino board

Next we will 'Upload to the I/O board' as shown in Figure 13: Upload to the I/O board, meaning we will secretly use AVRDude to send the hex code using the USB serial port to the bootloader on ATmega328 that will write it to memory. Though AVRDude is kept hidden, if you are curious, check out: http://savannah.nongnu.org/projects/avrdude/.

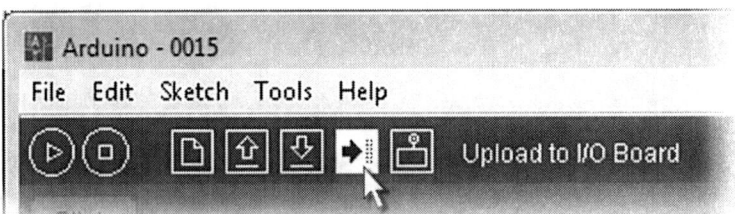

Figure 13: Upload to the I/O board

When you run this be sure and look at the TxD and RxD LEDs on the board right after you click the button so you can see them blinking rapidly in time to the flow of data between the Arduino board and the Arduino IDE. Sometimes it just helps to make things visible.

And after it loads, the Pin 13 LED should be blinking on and off once per second.

Getting Help

Before we get too deep into the Arduino, we should look at what – IMHO - is the best feature of the whole Arduino way of doing things: the **tools** that can help you understand and use it. There are two main tools. First is the website

30

Chapter 2: Arduino Quick Start Guide

www.arduino.cc - go there now and look around a bit. There are lots of things there that you may eventually find helpful. The second thing is the Help menu in the Arduino IDE. This item is seriously helpful. Click on it and browse around a bit to get a feel for what is there.

Using the Arduino Language Reference

We have already used some Arduino software without explaining how it works. You may remember in the Blink example that we had this line:

```
pinMode(ledPin, OUTPUT);
```

The pinMode() function is part of the Arduino language and you can find details about it by clicking on the Help\Reference item as shown in Figure 14: Help\Reference.

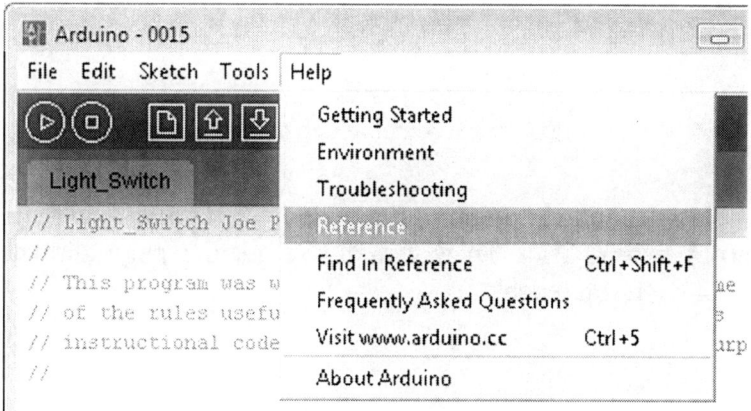

Figure 14: Help\Reference

This will cause the Arduino Language Reference to open in a browser as shown in Figure 15: Arduino Language Reference.

Chapter 2: Arduino Quick Start Guide

Figure 15: Arduino Language Reference

In the right column under Functions - Digital I/O you will see pinMode(pin,mode). Click on that and you get the page shown in Figure 16: pinMode function reference.

Chapter 2: Arduino Quick Start Guide

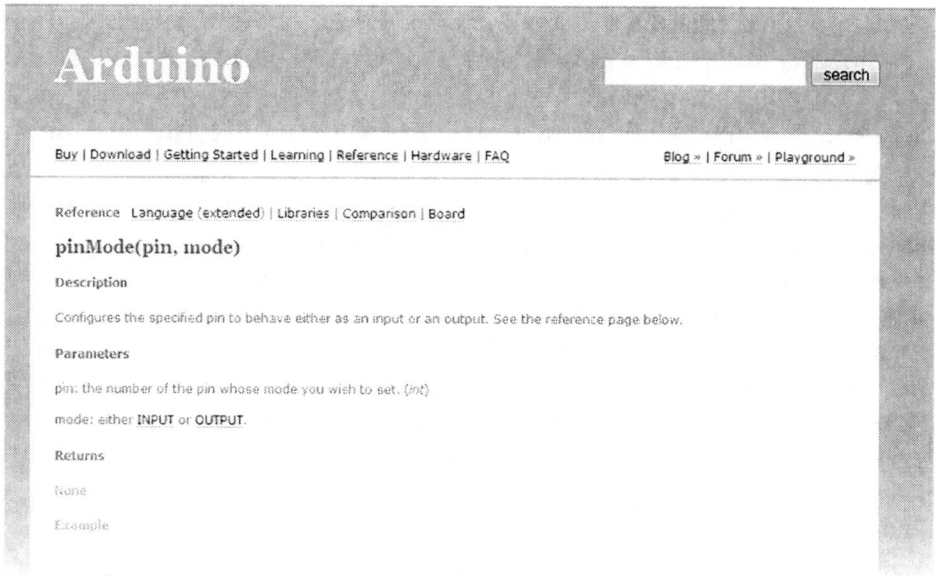

Figure 16: pinMode function reference

The discussions in this text should help familiarize you with many of the concepts, but the Help menu item is the ultimate resource.

Using Internet Forums

Another great thing about the Arduino is the forum on their website where you can ask questions. But please before you go to the forum and ask for help, make sure you have read all the introductory materials, tried to find the answer to your question in the Arduino Language Reference, and thoroughly Googled your problem. And your life will be considerably easier if before you start using any Internet forum you read: http://www.catb.org/~esr/faqs/smart-questions.html. Once you've exhausted all available resources, then please don't hesitate to ask your question on the forum.

Yes, you will get flamed

You will find many very helpful folks on the Internet, and you'll find some that are just plain bat-poop crazy and evil. I am not referring to the Arduino forum in

Chapter 2: Arduino Quick Start Guide

particular, since it seems to be a friendly place, especially when compared to some forums, but I just want to warn you that you need to disconnect your ego before using this resource. I have several friends who are robot enthusiasts who all agreed that they won't use my favorite forum (not the Arduino forum, and I won't name names here) because they are afraid of getting flamed. I had to bite my tongue to keep from saying, "Well, just hitch up your big-girl panties and get used to it. You want help on the Internet? You might get flamed and you might get help. But you sure as heck won't get help trembling in fear at the keyboard." There are folks on every forum who like to flame people. I even get so exasperated that I will occasionally get out my flame-thrower. But the thing to remember is that nobody actually knows you. You are no more a real person on the Internet, to some people, than a random character in a video game. Google 'Internet disinhibition' if this topic really interests you, but the take-home point is that some folks on Internet forums can be vicious while other's are nearly saints in there helpfulness, but none of it is aimed at you personally.

Chapter 3: Playing with software – Part 1

How do we write a program?

In the last chapter, we started playing with the Arduino and defined some terms. We defined a sketch (program) as a sequence of human readable text statements following the rules of a programming language that can be converted and uploaded to a micro providing instructions as to how it will make decisions about using sensors and actuators. While you will see the term 'sketch' frequently in Arduino discussions, from this point forward I will be using 'program' for that 'human readable text'.

Since this discussion is novice oriented, I'm going to repeat some stuff so bear with me. Software may be thought of as the instructions that tell a micro how to do things. Telling a computer what to do is called programming. We write a software program using tools that allow us to provide these computer instructions in some human comprehensible form, like text files that use words and symbols available on a keyboard. Our text file (source code) must follow a set of rules that are provided for a specific programming language, and then we must submit that file to some compiler software that will then do one of two things: compile it or spit it back in your face. The spitting part is accompanied with a list of complaints that provide a guide to making the source file more palatable. There are two kinds of complaints: warnings and errors. Errors will prevent the code from compiling, but warnings won't stop the compilation process, since the compiler will assume that you might just know what you are doing in the part that it is warning you about. You might be tempted to bypass warnings, but in general that is a very foolish thing to do.

If you didn't make any mistakes in the text file, the compiler will translate your source code into instructions that the computer can use. If you did make a mistake, the compiler will try to give you a hint of the problem. Sometimes these hints are useful; sometimes they are not.

So a program is some text that follows some rules and can be converted into something that we can feed the micro that it can use to make decisions. Text, rules, and decisions – we learn the rules, then we write the text, then the micro makes decisions. This is the sequence:

Chapter 3: Playing with software – Part 1

1. We decide what we want the micro to do.
2. We think about how our programming language rules might be used to get the micro to do our bidding.
3. We write a program using the text editor in the Arduino IDE (Integrated Development Environment).
4. The Arduino IDE sends that text to a compiler that checks our text against some rules and then builds something that can be uploaded to the micro. The micro accepts the upload and starts making decisions about how the Arduino board uses the hardware.

In the real world we make mistakes at each point in the sequence and we have to rework each step (debugging) until eventually we get it right.

Following this list means that in order for us to write a program we are going to have to learn two things. What are the rules we can use and what are the decisions the micro can make? Fortunately we don't have to learn it all at once before we get started playing with these things. We will learn bits and pieces in small steps.

How an Arduino Program is structured

An Arduino program is structured in four parts.

FIRST: Begin with some comments about the program

SECOND: List variables and constants that all the functions may use. Variables are names for memory locations that a computer can use to store information that might change. Constants are numbers that won't change.

THIRD: Run the setup() function to prepare stuff. This is where you perform tasks that you want done once at the beginning of your program

```
void setup()
{
   // do things once at the start of the program
}
```

Chapter 3: Playing with software – Part 1

FOURTH: Run the loop() function. This is where you run things in a sequence from the top of the loop to the bottom of the loop, then you start over again at the top, looping until the machine gets turned off

```
void loop()
{
  // Do the first thing

  // Do the second thing

  // Do any number of things

  // Do the last thing in the list

  // Go back to the beginning of this list
}
```

Structure of the Arduino Blink example

I have taken the Blink program from the Arduino Examples and added the lines showing each of the four sections.

FIRST: Opening comments:
```
/*
 * Blink
 *
 * The basic Arduino example.  Turns on an LED on for one second,
 * then off for one second, and so on... We use pin 13 because,
 * depending on your Arduino board, it has either a built-in LED
 * or a built-in resistor so that you need only an LED.
 *
 * http://www.arduino.cc/en/Tutorial/Blink
 */
```

SECOND: List variables
```
int ledPin = 13;   // LED connected to digital pin 13
```

THIRD: Things to do once at the beginning
```
void setup()        // run once, when the sketch starts
{
   pinMode(ledPin, OUTPUT);    // sets the digital pin as output
}
```

Chapter 3: Playing with software – Part 1

FOURTH: Do things in a loop
```
void loop()                            // run over and over again
{
  digitalWrite(ledPin, HIGH);    // sets the LED on
  delay(1000);                         // waits for a second
  digitalWrite(ledPin, LOW);     // sets the LED off
  delay(1000);                         // waits for a second
}
```

In the Quick Start Guide section we saw how to write a program in the Arduino IDE, compile it, and upload it to the Arduino board. You can locate this program in the Arduino IDE menu:

File\Sketchbook\Examples\Digital\Blink.

Click on the Blink then follow the instructions from the Quick Start Guide to compile and upload.

Learning the programming language rules

Arduino is, among other things, a programming language. But one of the first things to learn about the Arduino language is that underlying it is the C programming language. What Arduino does is to provide a novice-friendly buffer that hides some standard C requirements and adds some pre-written functions that help the novice use the Arduino board. You don't need to know a thing about C in order to use Arduino or this book. There is an Arduino way of doing things, and that is what we will learn in this section. I just wanted to let you know that most of the stuff we learn here is applicable to C and provides IMHO a good introduction to C if the reader chooses to move up (IMHO) to C.

We will write a simple program and then dissect it to study some of the rules.

Chapter 3: Playing with software – Part 1

An Arduino Light Switch

What do we want our program to do

In the Arduino Quick Start section we used the Arduino Blink example to get an LED blinking on the Arduino board. Some of what we did might even make a bit of sense without having the details explained. Now we are going to add to that program and read a pushbutton switch, and use that reading to turn an LED on or off. Yup, we are going to make an overly complicated and way too expensive light switch. Just remember, our goal at this point is to learn about software, not to create hardware devices that make a lot of sense.

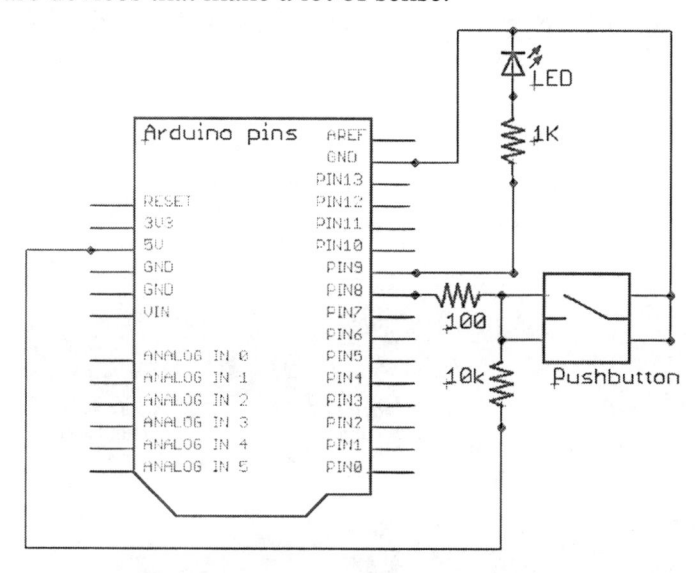

Figure 17: Schematic – Pushbutton switch and LED

In hardware we could directly turn the LED on or off with the switch, but since it is a momentary pushbutton switch, it only conducts electricity when pressed. As soon as you release it, the light will go off. But we want to have the button turn the LED on if it is off and off if it is on each time the button is pushed. To do this the micro will need to remember the LED state (is it on or off) and toggle that state each time the button is pushed. (Toggle means that if something is on, turn it off, and if it is off, turn it on.) And since we've got our hardware connected to a PC, we might as well add a couple of features, one to allow us to keep count of the number of times the button has been pressed and another that allows us to turn

39

Chapter 3: Playing with software – Part 1

the LED on or off from the PC keyboard. These extras will provide extra examples on how to apply the programming rules.

We want this program to:
1. Keep the LED status in memory in a variable named onOff that can be set to either true for on or false for off.
2. Read the status (on or off) of a pushbutton switch.
3. Toggle the value of the onOff variable each time the button is pushed.
4. Turn an LED on or off depending on the value in an onOff variable.

Set up the hardware

Refer to Figure 17: Schematic – Pushbutton switch and LED, and Figure 18: Drawing - pushbutton switch and LED, and add the parts to the breadboard.

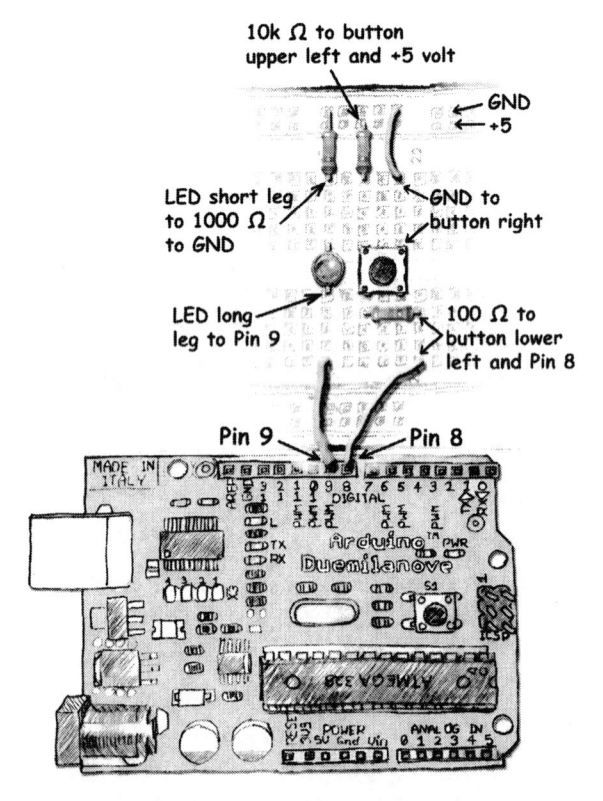

Figure 18: Drawing - pushbutton switch and LED

40

Chapter 3: Playing with software – Part 1

Now type the following program into the Arduino IDE.

Light_Switch - source code:

```
/* Light_Switch Joe Pardue 9/16/09
 *
 * This program was written for novices to demonstrate some
 * of the rules useful for programming with Arduino. It is
 * Instructional code and is not intended for any other purpose.
 */

//
//define constants
//
// pin for LED
#define LEDPIN 9
// pin for pushbutton
#define BUTTONPIN 8
// maximum count to assure the button was really pressed
#define DEBOUNCE 10

//
// define and initialize variables
//
// read button status
int reading = 0;
// count button detected LOW
int count = 0;
// track if LED is on (true) or off (false)
boolean ledState = false;

void setup()
{
  // set button pin to input
  pinMode(BUTTONPIN, INPUT);
  // set LED pin to output
  pinMode(LEDPIN, OUTPUT);
}

void loop()
{
  // wait for a valid button press
  while(count < DEBOUNCE)
  {
     checkButton();
  }
```

Chapter 3: Playing with software – Part 1

```
    // reset count
    count = 0;

    // toggle LED state
    if(ledState == true)
    {
        // if it was on (true) turn it off (false)
        digitalWrite(LEDPIN, LOW);
        ledState = false;
    }
    else
    {
        // if it was off (false) turn it on (true)
        digitalWrite(LEDPIN, HIGH);
        ledState = true;
    }

    // give us time to get our finger off the button
    delay(500);

}

// Check the button and count up if it is connected to
// ground (LOW) or count down if it is connected to +5 (HIGH)
void checkButton()
{
  // read the pin HIGH or LOW?
  reading = digitalRead(BUTTONPIN);

  // if it is LOW and the count is less than DEBOUNCE
  // increase the count by 1.
  if ( (reading == LOW) && (count < DEBOUNCE) )
  {
    count++;
  }
  else// otherwise if it is greater than 0, decrease by one
  {
    if(count > 0) count--;
  }
}
```

If you are really new at this you likely noticed some things in that program that haven't been explained yet, well don't panic they will be explained eventually and

Chapter 3: Playing with software – Part 1

if you wait to write your first program until you understand everything, you'll never write that first program.

I strongly recommend that you type each and every word of this program into the Arduino IDE. While you can get the source code from my website and copy and paste it, you will miss a lot of very important education, not just in getting to carefully look at each character as you type it in, but you will miss out on the debugging process that will result from inevitable typos. Remember debugging is not an optional chore, it is necessary and if you are the type of person who ultimately succeeds, debugging is kind of fun.

Some of the rules

This section takes a very brief look at the Light_Switch program to help you understand what each line means.

Comments

You can add comments (text the compiler ignores) to your code two ways.
For a single line of comments use double back slashes as in:

```
// give us time to get our finger off the button
delay(500);
```

For multi-line comments, you can begin them with /* and end them with */ as in:

```
/*
  Light_Switch Joe Pardue 9/16/09

  This program was written for novices to demonstrate some
  of the rules useful for programming with Arduino. It is
  Instructional code and is not intended for any other purpose.
*/
```

Functions

A function encapsulates a computation. Think of them as building material for programs. A house might be built of studs, nails, and panels. The architect is assured that all 2x4 studs are the same, as are each of the nails and each of the panels, so there is no need to worry about how to make a 2x4 or a nail or a panel,

Chapter 3: Playing with software – Part 1

you just stick them where needed and don't worry how they were made. In the Light_Switch program, the setup() function twice uses the pinMode() function:

```
// set button pin to input
pinMode(BUTTONPIN, INPUT);
```

The writer of the setup() function doesn't need to know **how** the pinMode function does its job, he only needs to know **what** it does and what parameters to use. In this case the first parameter is the Arduino pin name: BUTTONPIN and the second is the mode for using that pin: INPUT. Later, you may want to learn a bit more about the function pinMode() and you can find out how to use it in the Arduino library reference. Even later you may want to learn how it works and you can find it in the library source code. But the critical point is that NOW you can use this function and you don't need to know how it works. It is like the above example of a wood 2x4 used in construction; you don't need to understand how a tree makes wood to build a house.

Encapsulation of code in functions is a key idea in programming and helps make chunks of code more convenient to use. And just as important, it provides a way to make tested code reusable without having to rewrite it for each use.

You are <u>required</u> to use the setup() and loop() functions in an Arduino program. You are <u>allowed</u> to use the prewritten functions such as pinMode() discussed in the Language Reference. And you can build your own functions like the checkButton function in the Light_Switch program.

When you build your own function such as checkButton(), the source code begins with a line that has the data type that the function returns when it is called by another function (the data type is void if it doesn't return anything). The data type is followed by the function name, then a parameter list enclosed by parenthesis. (The checkButton example doesn't return anything (so it is the void type) and doesn't require any parameters) The pinMode function required two parameters. The parameter list consists of variables that are provided to the function and used in it. The parameter list is followed by an open curly bracket: '{' and a closed curly bracket: '}' between which you will find the things that the function does.

Expressions, Statements, and Blocks

Expressions are combinations of variables, operators, and function calls that produce a single value. For example:

```
reading = digitalRead(BUTTONPIN)
```

This is an expression that sets the 'reading' variable to the value returned by the digitalRead(BUTTONPIN) function.

Statements control the program flow and consist of keywords, expressions, and other statements. A semicolon ends a statement. So for example we add a semicolon to our expression and we get a statement:

```
reading = digitalRead(BUTTONPIN);
```

Blocks are compound statements grouped by open and closed curly brackets: { }. For example:

```
if(ledState == true)
{
    // if it was on (true) turn it off (false)
    digitalWrite(LEDPIN, LOW);
    ledState = false;
}
```

This groups the two inner statements to be run depending on the condition of the 'if' statement. Brackets are also error prone and they can be hard to follow in complex code. Just get used to the process of matching brackets as part of your debugging routine.

Flow Control

Flow control statements dictate the order in which a series of actions are performed. For example in the loop() function we use 'while' to keep us in an inner loop until the count variable (which is incremented in the checkButton() function) becomes equal to (or greater than) the DEBOUNCE constant:

Chapter 3: Playing with software – Part 1

```
// wait for a valid button press
  while(count < DEBOUNCE)
  {
     checkButton();
  }
```

After running the checkButton() function, we exit the while loop and then set the count back to zero . Next we use the 'if...else' to decide to turn the LED on or off:

```
// reset count
count = 0;

// toggle LED state
if(ledState == true)
{
    // if it was on (true) turn it off (false)
    digitalWrite(LEDPIN, LOW);
    ledState = false;
}
else
{
    // if it was off (false) turn it on (true)
    digitalWrite(LEDPIN, HIGH);
    ledState = true;
}
```

If you aren't 100% clear on all this yet, don't panic, we'll revisit all these ideas as we move along. You may be especially confused by some of the symbols such as '>' or '==' and well you should be. These things are called operators and will be discussed in more detail in Chapter 7: Playing with software – Part 2.

Arduino Functions

In the Arduino Reference you will see a list of functions that allow you to use the Arduino board resources such as reading and writing to digital and analog pins. We will see how to use many of these functions in the projects section where they will be applied to playing with the hardware in the Smiley Micros Arduino Projects Kit.

Chapter 4: Hardware Prototyping

Introduction to Breadboards

I tend to write this stuff with the idea that my reader already knows a bit about electronics. But that isn't always a valid assumption. For instance, some folks have never used a solderless breadboard. I tried to remember the first time I used one and it just seems like I was born knowing how these things work, but I do have vague recollections of using a ohm meter to figure out that the + and − power busses run horizontal the entire length of the board and that the two grids in the middle of the board have vertical 5-position clips. If you have already used one of these things then skip over this section.

How a breadboard works

In the good old days, electronics experimenters would build prototypes by nailing components to an actual wooden breadboard and then soldering wire between connection points. Today's solderless breadboards are made of plastic blocks with holes on 0.1-inch centers that allow you to insert jumper wires (usually 22 AWG) into hidden clips below the holes. The vertical 'terminal' blocks let you connect up to 5 points on each of the 63 double sets of columns. These columns are separated by a 0.4-inch gutter over which you can place an IC or DIP package. You also have 4 horizontal power bus rows with 50 point clips on each… **oh who am I kidding**, there is no way you can really understand what is going on by reading a description of this. It is even hard to take decent photographs and have it clear what I'm are talking about, so I decided to take some photos and get out the crayons and draw some pretty pictures that might just make things clearer. Figure 19 shows the top and bottom of a solderless breadboard (the bottom has the foam tape stripped off to show the connections. Figure 20 shows the clips pulled out. Figure 21 shows how a clip grabs a wire and Figure 22 shows a cutaway drawing with an LED, 1K Ω resistor, and a jumper wire all connected such that if you have +5 volts in the upper + channel and GND in the lower − channel, the LED should light up.

Chapter 4: Hardware Prototyping

Top view of breadboard.

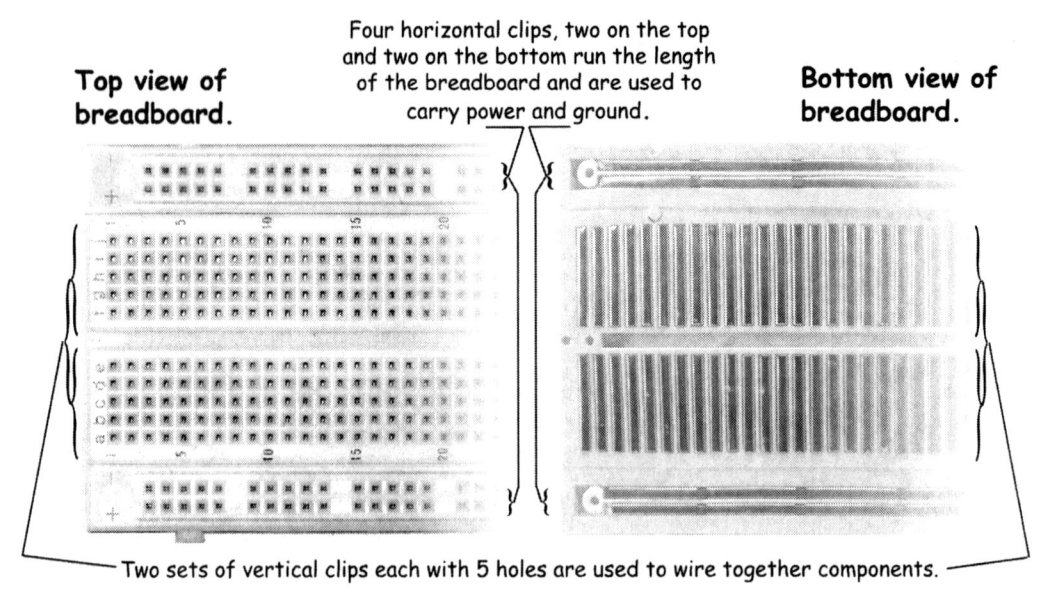

Four horizontal clips, two on the top and two on the bottom run the length of the breadboard and are used to carry power and ground.

Bottom view of breadboard.

Two sets of vertical clips each with 5 holes are used to wire together components.

Figure 19: Breadboard front and back

Bottom view of breadboard with vertical

Bottom view of breadboard with horizontal power clip pulled out.

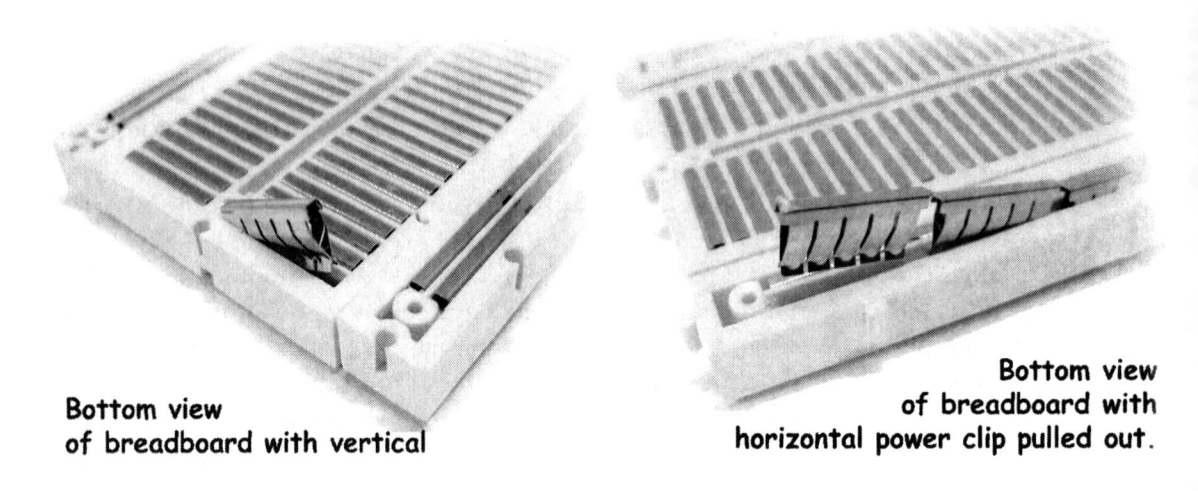

Figure 20: Back with 5-position vertical and 50-position clips

Chapter 4: Hardware Prototyping

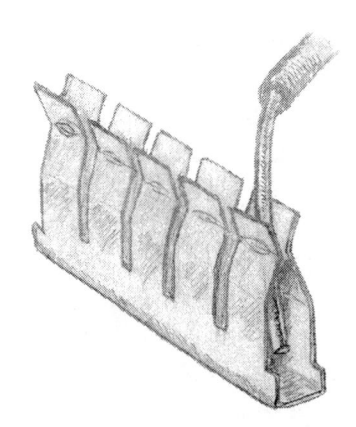

Figure 21: A 5-position clip

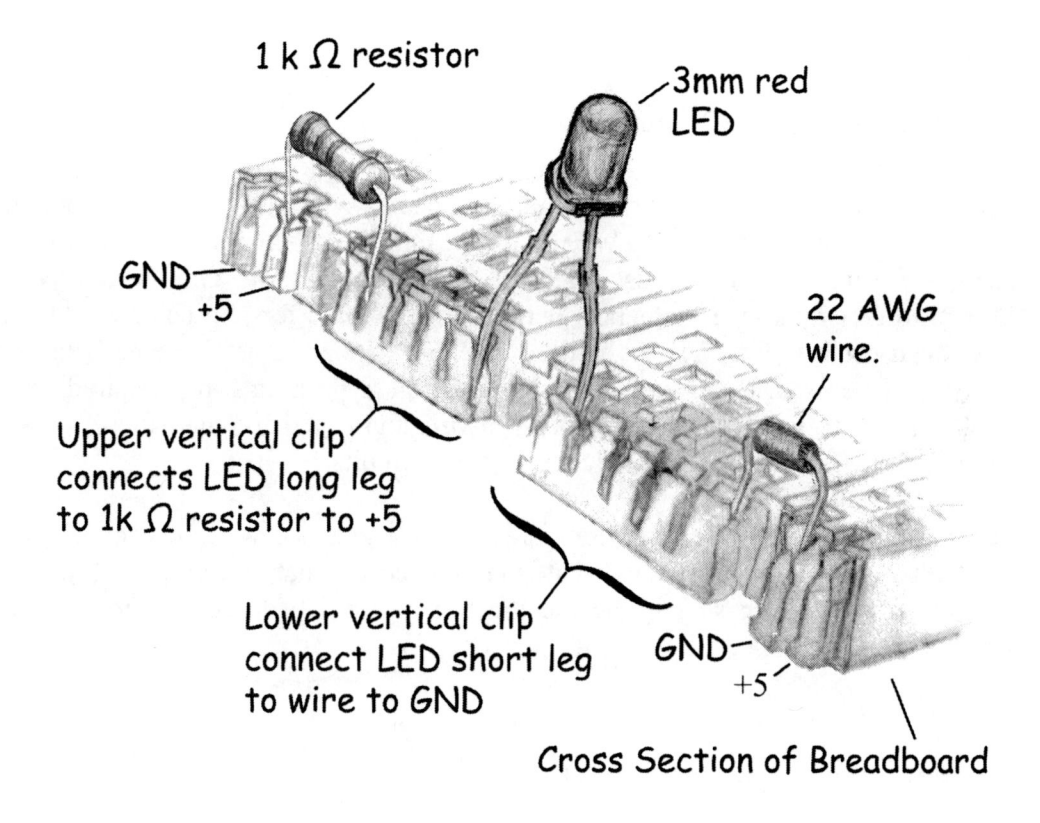

Figure 22: Breadboard cross-section with LED, resistor, and wire

Chapter 4: Hardware Prototyping

The clips should handle about 1 amp at 5-volts. Breadboards only work for relatively low frequency devices (10MHz or less) [due to high and variable stray capacitance and inductance – concepts we'll defer for another book]. Also, the jumpers don't always maintain a solid connection. I can't count the times that I've had a circuit go weird until I jiggled a few wires and the weirdness went away. However, I've gotten microcontrollers to work with 16 MHz external clocks on breadboards, but it isn't something that you can be totally confident in, so caution is advised. You are trading reliability for flexibility (and price). You will sometimes see an 'Arduino' design with the ATmega and associated circuits on a breadboard, and there is nothing really wrong with doing that, but every jumper tie point just adds another possible place for a bug, so you have to ask yourself if your time is worth the hassle when you can get the Arduino on a PCB pretty cheap.

An Introduction to Schematics

Schematics are drawings of circuits that use special symbols for each component and show connections as lines. We usually design the schematics for our circuits using software on a PC. I use EAGLE (Easily Applicable Graphical Layout Editor) to generate our schematics. It has a free version for non-commercial use (www.cadsoft.de). EAGLE is hard to use, as are all other schematic/layout programs I've used. If you want to use any of these programs, be prepared for a long learning curve. Figure 23: Layout Drawing and EAGLE Schematic for LED, shows a drawing and the schematic for an LED circuit.

Schematic connections can get confusing to look at since wires can cross over each other. The software knows what is connected to what, but you can't always tell just from looking so I'll use the junction symbol (a large dot) to indicate where wires are connected.

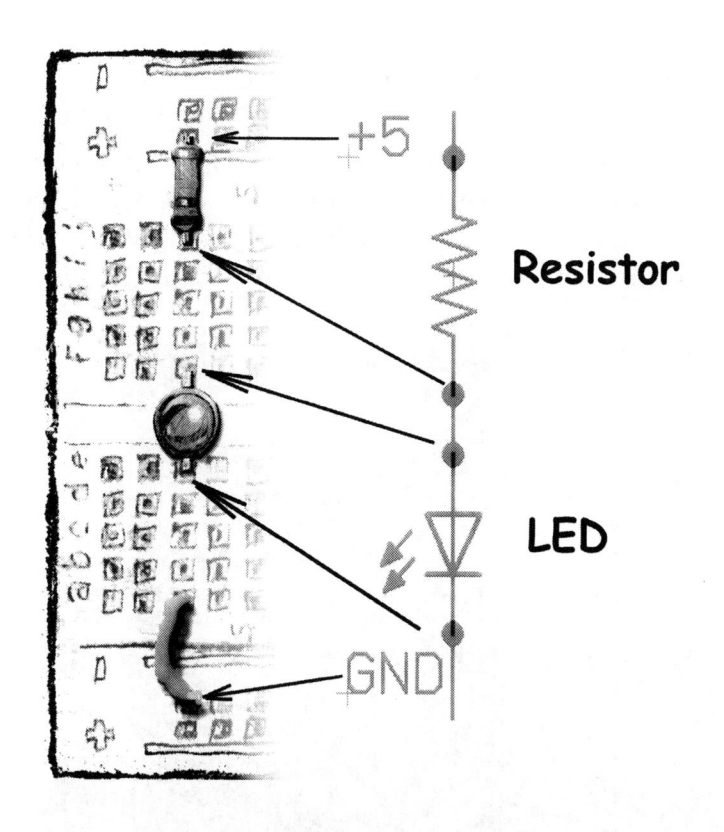

Chapter 4: Hardware Prototyping

Figure 23: Layout Drawing and EAGLE Schematic for LED

Schematic Symbols

Schematic symbols are not standardized, but you often see symbols similar to the ones shown in Figure 23: Layout Drawing and EAGLE Schematic for LED. A resistor is often shown in America with a ziz-zag, but in Europe they may use a rectangle. The LED symbol is also the symbol for a diode, but with a couple of arrows added to show light coming out (LED is the acronym for Light Emitting Diode). A diode is a component that allows conventional current to move in one direction as shown by the triangle and blocked in the other direction as shown by the bar at the end of the arrow. We will see other symbols as we build more circuits with the Arduino Projects Kit components.

Chapter 4: Hardware Prototyping

Using a breadboard with the Arduino

Figure 24: Arduino pin-out schematic shows a schematic symbol for the Arduino along with the drawing of the Arduino that shows the female headers associated with the pin numbers. These headers have clips in them similar to those on a breadboard, except that there is only one clip per hole.

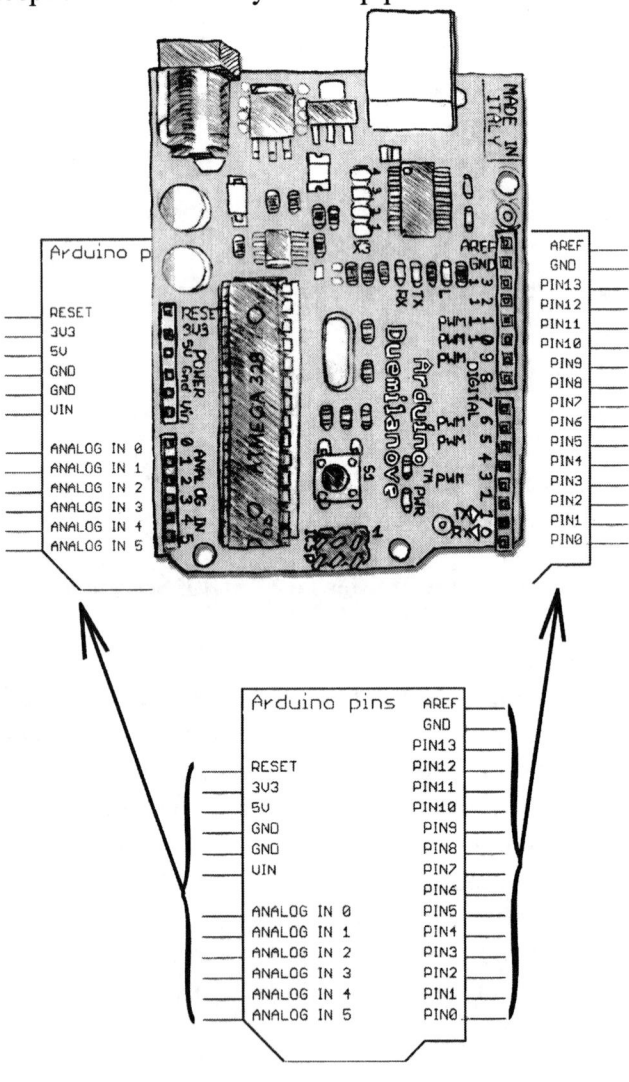

Figure 24: Arduino pin-out schematic

Chapter 4: Hardware Prototyping

The Arduino Learning Platform (ALP)

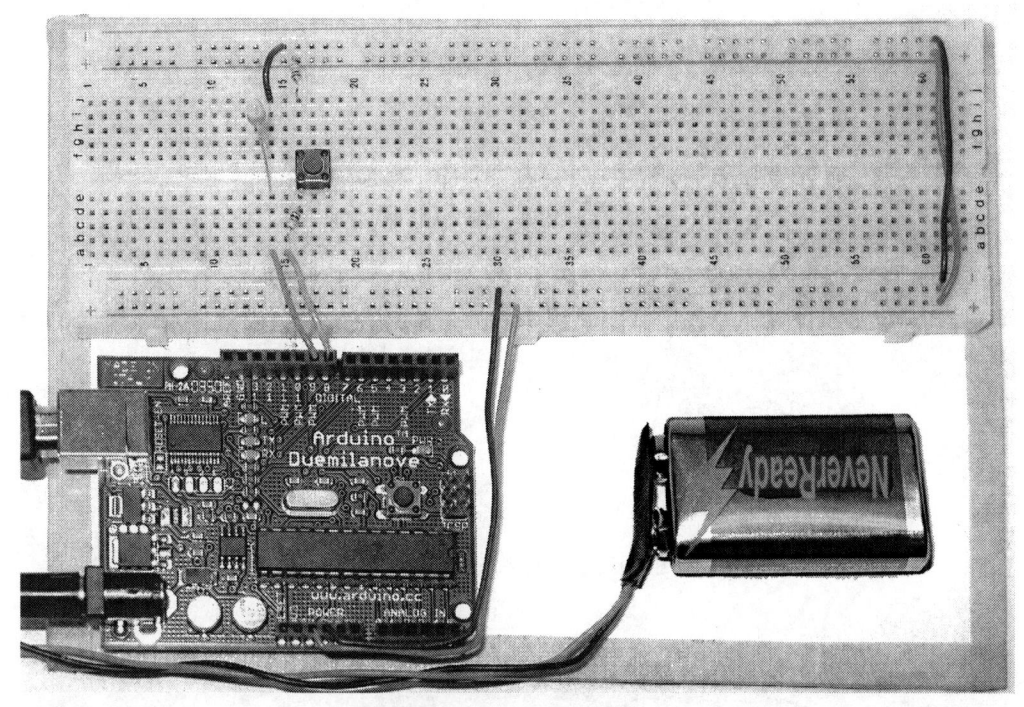

Figure 25: The ALP base.

Let's build a protective environment for our experiments. The Arduino Learning Platform (ALP) consists of an Arduino, Breadboard, and 9-volt battery mounted on a foamcore board that slides into a groove (Figure 26: Slide it in the ALP box) in a box to protect the breadboard circuits from getting knocked about.

I keep the box closed with a rubber band that also holds the USB cable. I've taken this box out hundreds of times stuffed in a backpack along with my laptop and worked in various clean well-lighted places while sipping too-expensive tea and haven't damaged this little system or even dislodged any of the wires on the breadboard prototypes. Note though in Figure 27: Scruffy protective box, that it is getting a little dinged up and marked. I am personally amazed that it has held up and that the masking tape makes such a good hinge that hasn't ripped yet. Sure, its nerdy and cheap, but so am I.

Chapter 4: Hardware Prototyping

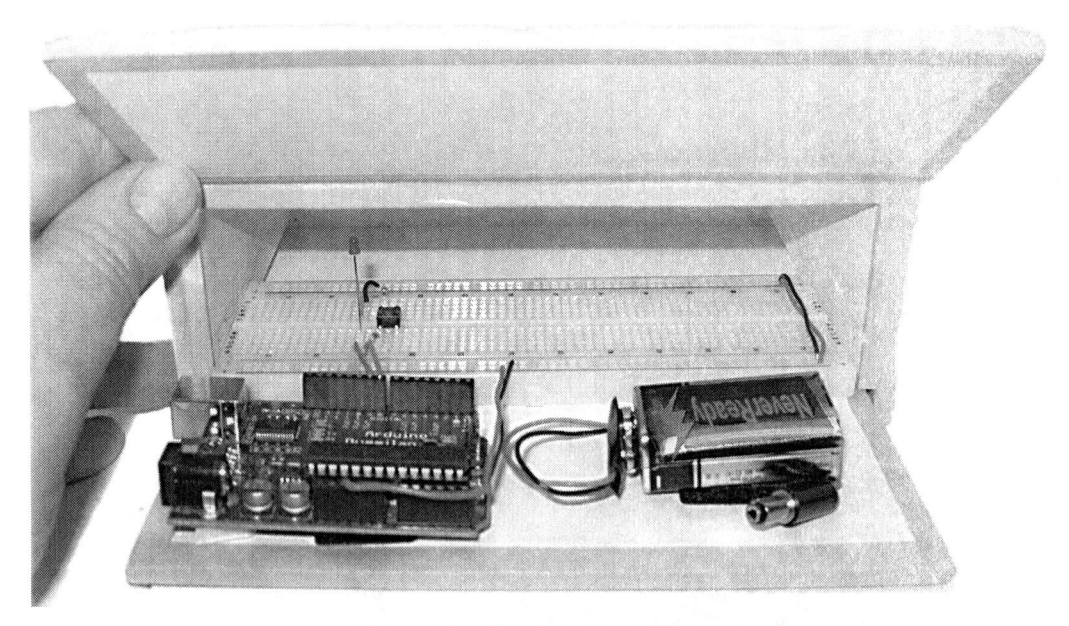

Figure 26: Slide it in the ALP box

Figure 27: Scruffy protective box

Chapter 4: Hardware Prototyping

Figure 28: Construction site

You could use corrugated boxboard, but the foam core board is a bit sturdier (and prettier). You'll need a razor sharp knife (a box cutter with break-off tips is a good choice), white glue, masking tape and a cutting surface that you don't mind accidentally cutting. For mounting the Arduino and battery box you will need foam tape and Velcro (also patience and luck). A nice first aid kit might also come in handy.

Cut out foamcore panels:
Arduino Breadboard base 7" x 4 3/4"
Box Base – 7 1/16" x 4 13/16"
Back – 7 3/8" x 2 1/2"
Outer sides – 4 13/16" x 2 1/2"
Inner sides – 4 13/16" x 2"
Top – 7 3/8" x 5"
Front – 7 7/16" x 2 11/16"

ALP Base

In Figure 29: Stick on Velcro squares, we add a some Velcro to the Arduino, battery, and board so that we can keep them in place except when we need to do things like replace the battery or remove the Arduino for some other use. The

Chapter 4: Hardware Prototyping

breadboard comes with a foam tape base that you can stick directly to the board, or you can use Velcro on it if you think you might later need it somewhere else.

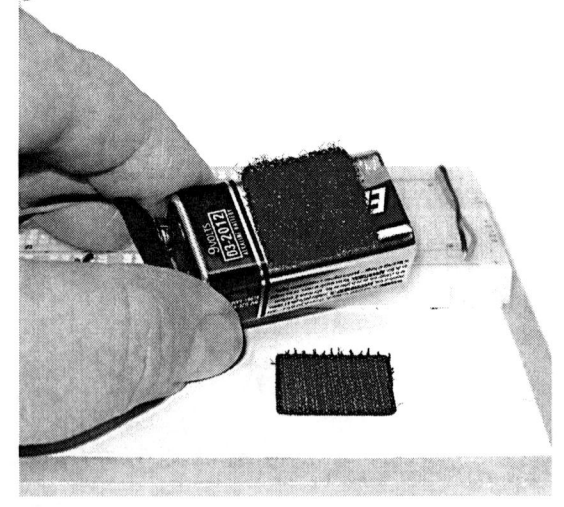

Figure 29: Stick on Velcro squares

Arduino Learning Platform Storage Box

Figure 30: Start assembly

Chapter 4: Hardware Prototyping

Notice (Figure 30: Start assembly) that the sides have a tall and a short panel glued together. This will provide a slot for the development board to slide into (as shown in Figure 26: Slide it in the ALP box). Be careful when making this slot so that it is big enough and doesn't have a lot of glue and crap to inhibit sliding the baseboard in.

Figure 31: Glue the sides together as shown

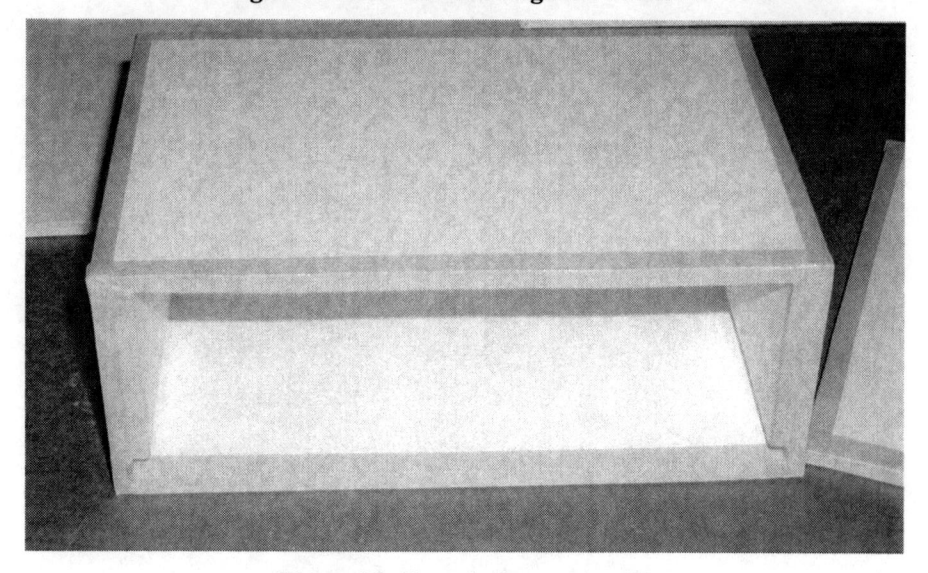

Figure 32: Front before taping lid

57

Chapter 4: Hardware Prototyping

Cover the box edges with masking tape being careful with the slot that will hold the baseboard. Cover the lid edges with masking tape and then tape the lid to the box using tape on both sides to form a tape hinge.

Slide the baseboard into the box. Close the lid and secure it with a rubber band.

Now admit it: that was fun, wasn't it?

Chapter 5: Some Simple Projects

Chapter 5: Some Simple Projects

In this chapter we are going to apply some of what we've learned to do some simple projects. First we will put a DIP switch and LED on our breadboard and learn how to read the switch states and set the LED states. Next we will make a simple speaker and generate some 'music' (you'll hear why I use the quotes when you get this project squawking). And finally we'll make an LED fade off and on.

Digital Input and Output

DIP Switch and LEDs

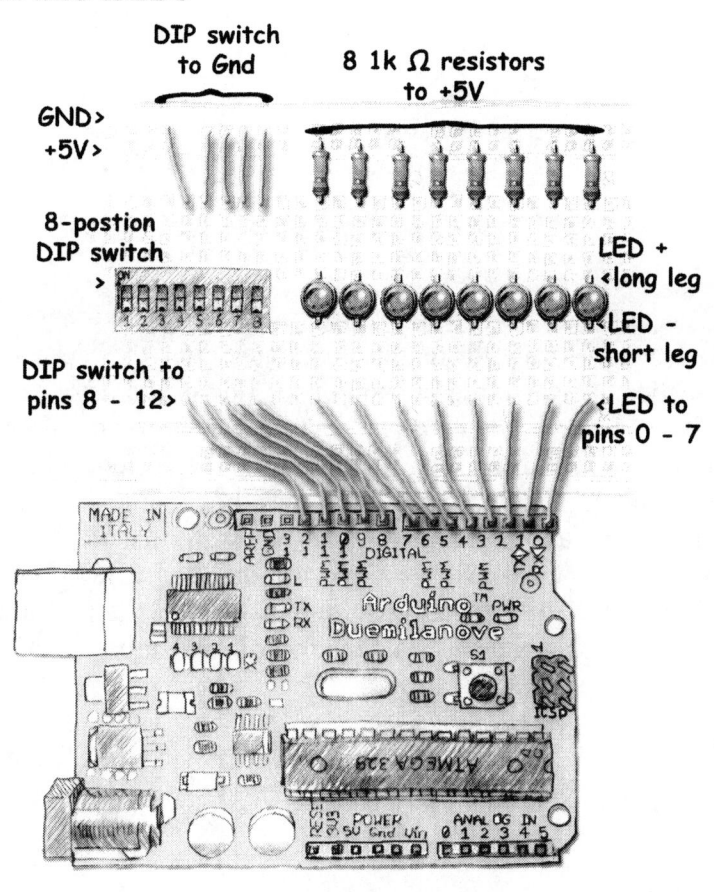

Figure 33: DIP switch and LEDs drawing

Chapter 5: Some Simple Projects

In this section we will be do some digital I/O (Input/Output) using an 8-position DIP switch for input and 8 LEDs for output. We will first learn how to read the switches and show that reading on the LEDs.

The DIP Switch

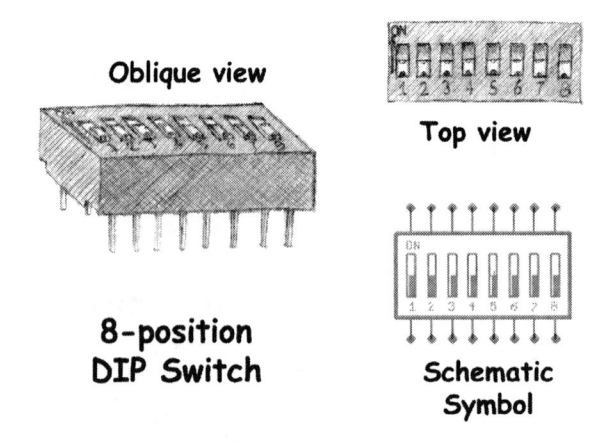

Figure 34: 8-position DIP switch.

DIP switches get their name from Dual Inline Package and they were designed to fit in sockets or pads made for regular integrated circuits. These switches make or break connections between the pins on either side of the device depending on the position of the switch in the center. They were designed to allow users to preset values that a computer could read for setting up the system. This means that they were made to be used a relatively few times, not frequently like a light switch. You shouldn't have any problem but just be aware that if you want to design a circuit where the switches will be thrown more than a few hundred times, you should probably look for something more robust. Also, the pins are very fragile so be careful when using them on a breadboard, and use a small screwdriver to remove them.

We will only use 5 of the DIP switch's 8 pins since we want to reserve the remaining 8 Arduino digital I/O pins for the LEDs.

Chapter 5: Some Simple Projects

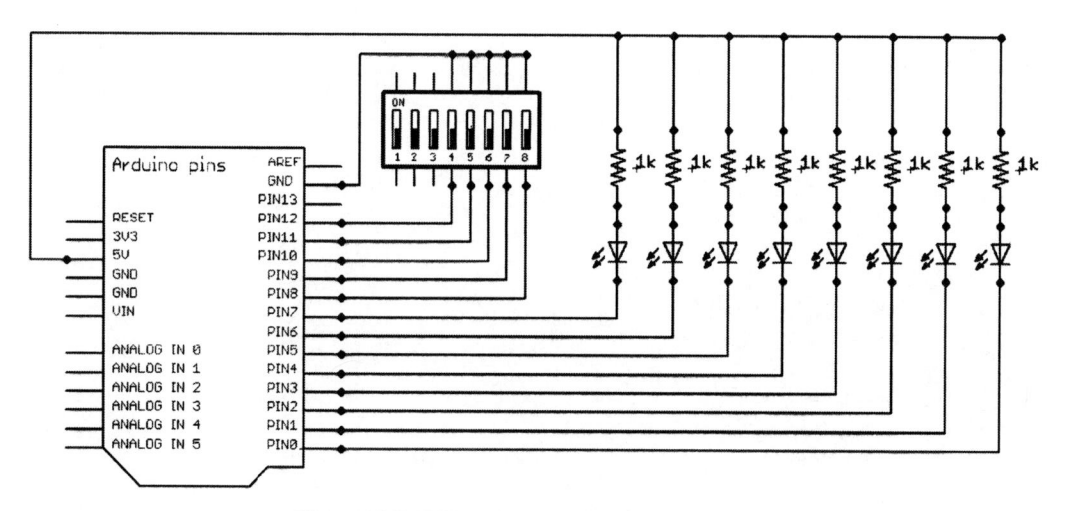

Figure 35: DIP switch and LED schematic.

DIP to LED Source Code:

```
// DIP to LED
// Joe Pardue September 29, 2009

void setup()
{
 // Init pins for input
 pinMode(8, INPUT); // DIP 0
 digitalWrite(8,HIGH); // Turn on pullup
 pinMode(9, INPUT); // DIP 1
 digitalWrite(9,HIGH); // Turn on pullup
 pinMode(10, INPUT); // DIP 2
 digitalWrite(10,HIGH); // Turn on pullup
 pinMode(11, INPUT); // DIP 3
 digitalWrite(11,HIGH); // Turn on pullup
 pinMode(12, INPUT); // DIP 4
 digitalWrite(12,HIGH); // Turn on pullup

 // Init pins for output
 pinMode(0, OUTPUT); // LED 0
 pinMode(1, OUTPUT); // LED 1
 pinMode(2, OUTPUT); // LED 2
 pinMode(3, OUTPUT); // LED 3
 pinMode(4, OUTPUT); // LED 4
 pinMode(5, OUTPUT); // LED 5
 pinMode(6, OUTPUT); // LED 6
 pinMode(7, OUTPUT); // LED 7
```

Chapter 5: Some Simple Projects

```
  // preset to turn LEDs off
  digitalWrite(0,HIGH);
  digitalWrite(1,HIGH);
  digitalWrite(2,HIGH);
  digitalWrite(3,HIGH);
  digitalWrite(4,HIGH);
  digitalWrite(5,HIGH);
  digitalWrite(6,HIGH);
  digitalWrite(7,HIGH);
}

void loop()
{
  digitalWrite(0,digitalRead(8));  // show DIP 0 state on LED 0
  digitalWrite(1,digitalRead(9));  // show DIP 1 state on LED 1
  digitalWrite(2,digitalRead(10)); // show DIP 2 state on LED 2
  digitalWrite(3,digitalRead(11)); // show DIP 3 state on LED 3
  digitalWrite(4,digitalRead(12)); // show DIP 4 state on LED 4
}
```

Run the code and then watch as the five rightmost LEDs reflect the state of the DIP switches. We will be using the DIP switch and LED circuit for several other projects later on, especially when we learn about binary numbers.

Cylon Eyes the Arduino Way

We will make the LEDs scroll back and forth like Cylon Eyes using the Arduino style of writing to individual pins. Later in the book we'll see another way to accomplish this task by using pins grouped as ports. But for now let's do it the long way. If you type this in, you'll find the cut/paste edit tool very handy.

Cylon Eyes 1 Source Code:

```
// Cylon Eyes 1
// Joe Pardue September 30, 2009
// This is a much-simplified novice-friendly approach that uses
// individual pins with the pinMode() and digitalWrite()
// functions.
//
// Later in the book (Chapter 7) we will take a deeper look into
// grouping pins as ports and using binary numbers and boolean
// logic.

// 0.1 seconds between LED on/off in sweep
```

Chapter 5: Some Simple Projects

```
int interval = 100;

void setup()
{
 // Init pins for output
 pinMode(0, OUTPUT); // LED 0
 pinMode(1, OUTPUT); // LED 1
 pinMode(2, OUTPUT); // LED 2
 pinMode(3, OUTPUT); // LED 3
 pinMode(4, OUTPUT); // LED 4
 pinMode(5, OUTPUT); // LED 5
 pinMode(6, OUTPUT); // LED 6
 pinMode(7, OUTPUT); // LED 7

 // preset to turn LEDs off
 digitalWrite(0,HIGH);
 digitalWrite(1,HIGH);
 digitalWrite(2,HIGH);
 digitalWrite(3,HIGH);
 digitalWrite(4,HIGH);
 digitalWrite(5,HIGH);
 digitalWrite(6,HIGH);
 digitalWrite(7,HIGH);

}

void loop()
{
  // The LEDs are connected to 5V with 1k ohm resistors
  // To turn them on the Arduino pin connects them to LOW
  // To turn them off the Arduino pin connects them to High

  // Scroll right to left

  // Turn off the last LED, turn on the next
  digitalWrite(0,HIGH);
  digitalWrite(1,LOW);
  delay(interval);

  // Turn off the last LED, turn on the next
  digitalWrite(1,HIGH);
  digitalWrite(2,LOW);
  delay(interval);

  // Turn off the last LED, turn on the next
  digitalWrite(2,HIGH);
```

Chapter 5: Some Simple Projects

```
digitalWrite(3,LOW);
delay(interval);

// Turn off the last LED, turn on the next
digitalWrite(3,HIGH);
digitalWrite(4,LOW);
delay(interval);

// Turn off the last LED, turn on the next
digitalWrite(4,HIGH);
digitalWrite(5,LOW);
delay(interval);

// Turn off the last LED, turn on the next
digitalWrite(5,HIGH);
digitalWrite(6,LOW);
delay(interval);

// Turn off the last LED, turn on the next
digitalWrite(6,HIGH);
digitalWrite(7,LOW);
delay(interval);

// Scroll left to right

// Turn off the last LED, turn on the next
digitalWrite(7,HIGH);
digitalWrite(6,LOW);
delay(interval);

// Turn off the last LED, turn on the next
digitalWrite(6,HIGH);
digitalWrite(5,LOW);
delay(interval);

// Turn off the last LED, turn on the next
digitalWrite(5,HIGH);
digitalWrite(4,LOW);
delay(interval);

// Turn off the last LED, turn on the next
digitalWrite(4,HIGH);
digitalWrite(3,LOW);
delay(interval);

// Turn off the last LED, turn on the next
```

Chapter 5: Some Simple Projects

```
    digitalWrite(3,HIGH);
    digitalWrite(2,LOW);
    delay(interval);

    // Turn off the last LED, turn on the next
    digitalWrite(2,HIGH);
    digitalWrite(1,LOW);
    delay(interval);

    // Turn off the last LED, turn on the next
    digitalWrite(1,HIGH);
    digitalWrite(0,LOW);
    delay(interval);
}
```

Well, that looks cool and invites you to mess with the code a bit to see how changing some of the numbers changes the output. For instance change 'interval' from 100 to 500 or 50 and see what happens. Play with this and there is no telling what you might learn. Also, don't pull the components out of the board just yet (unless you just have too) since we will see this design later.

Using the individual pin version of Cylon Eyes is simple to understand and if you used the cut/paste facility of the editor, it isn't even that difficult to type in. But this isn't the way a program like this would normally be done in a microcontroller. We would usually use the concept of an I/O port which bundles together I/O pins, usually in chunks of eight that can then be dealt with using the micros native 8-bit byte handling facilities. We are going to repeat the Cylon Eyes in the Chapter 7 where we will learn to do this with ports and binary numbers. Yes, it is conceptually a bit more challenging, but that is why I'm saving it for later. [Oh, and you'll learn what this means: 'There are exactly 10 types of people in the world, those that know binary and those that don't.']

Chapter 5: Some Simple Projects

Output sound – piezo music

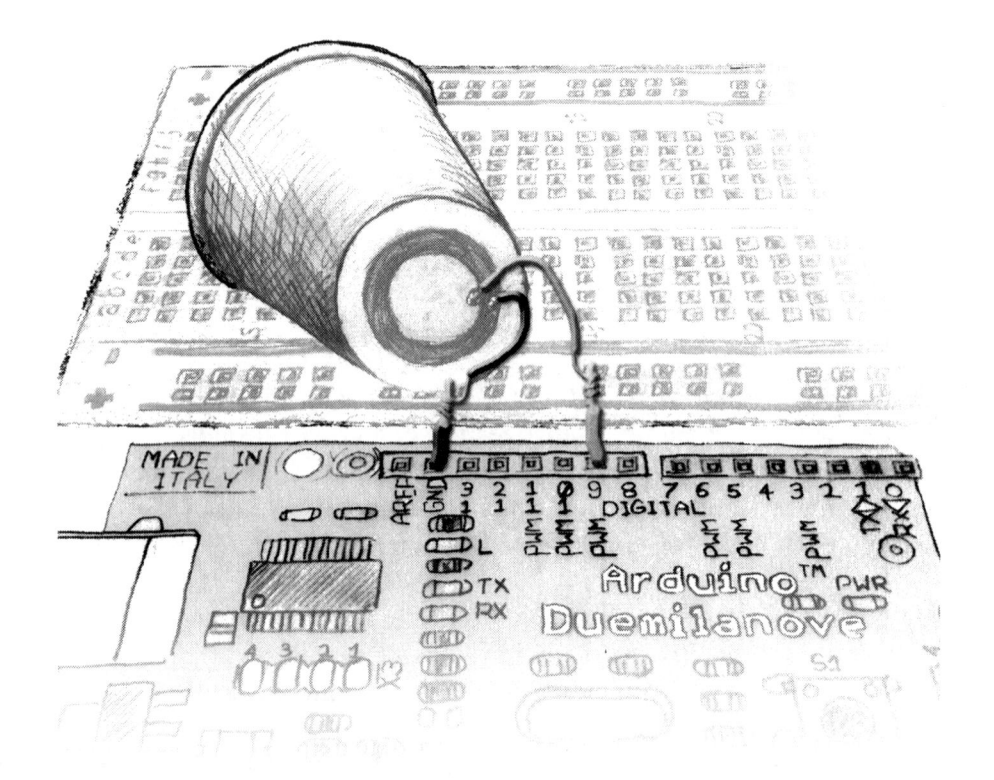

Figure 36: Piezo element layout.

We are advised to make a joyful noise, and what better than a piezo element? Well, honestly, just about anything would be better – you don't get much more low-fidelity than this. Even calling the sound it makes 'noise' is being generous - so let's ask 'what cheaper'? And now we are getting somewhere since these things are cheap and don't require any external amplification circuitry.

The Arduino example: 'Melody' by D. Cuartielles' plays 'Twinkle Twinkle Little Star' and I have modified that code here to play 'Happy Birthday' instead. (In Chapter 8: Communicating with a PC, section: 'Jukebox' Tunes Source Code, we will show how to play both tunes and some siren sounds, all selectable from the Serial Monitor.)

Chapter 5: Some Simple Projects

The piezo element warps in response to voltage changes and if this warping is at audible frequencies you can hear it. I glued (Elmer's©) the brass side of the piezo to the outside base of a Dixie© cup (one of those small cups you sometimes see in bathroom dispensers). You may be able to hear the sound without the cup, but the cup provides a resonant cavity (or some such techno-buzz words) that mechanically amplifies and directs the sound.

BTW, there are many piezo-buzzers out there and they often have special circuitry to create their own buzz, meaning they are either quiet or squalling, but can't be made to output a specific frequency and are not suitable for this project.

Sounds Components, Schematic, Layout

The piezo element in the Arduino Projects Kit is made from a brass disc with a ceramic disc adhered to it. The brass has the negative (black) wire and the ceramic has the positive (red) wire soldered to it as shown in Figure 36: Piezo element layout.

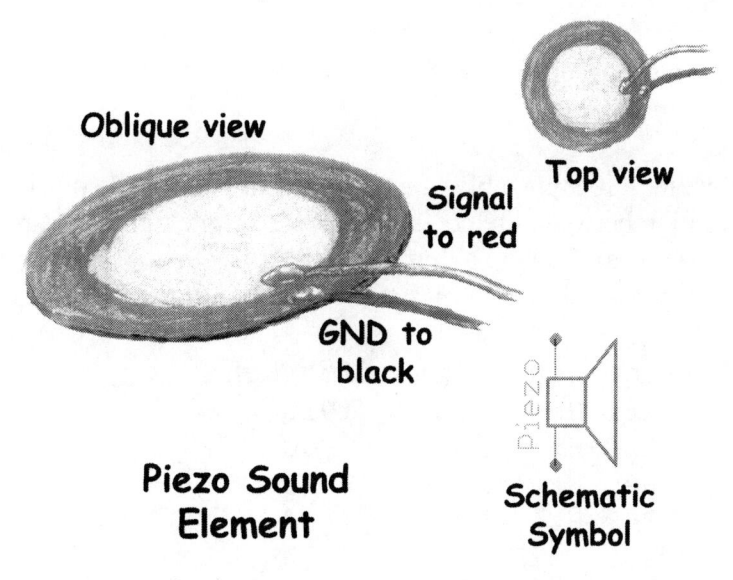

Figure 37: Piezo element illustration

These wires are stranded and can't be used directly with a breadboard, so take two pieces of 22 AWG solid wires (the kind that comes with the projects kit) and solder them as extensions to the piezo wires. See Figure 38: Piezo element

Chapter 5: Some Simple Projects

schematic. You don't even need a breadboard with this demonstration since you can plug the black wire directly into the Arduino GND and the red into the pin 9 socket as shown in Figure 36: Piezo element layout.

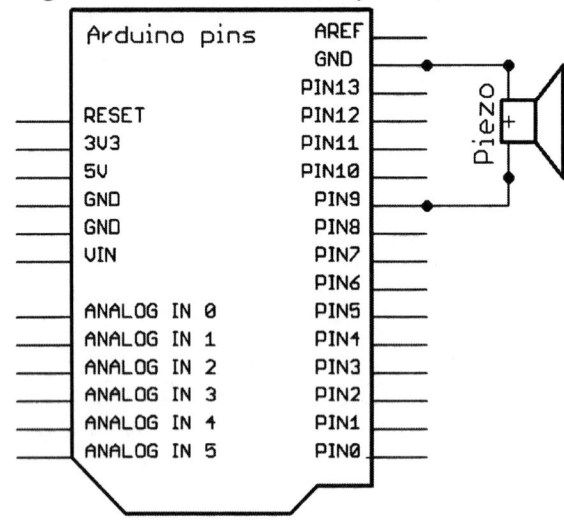

Figure 38: Piezo element schematic.

Tunes

For us to create a recognizable tune we need to control the musical notes (tones) and the duration between the notes (beat). For simple tunes we can live with eight tones (a music octave) each having a specific a frequency. Each of these tones has a letter 'note' assigned to it by musicians as in Figure 39: Note table.

```
* note  frequency  period  timeHigh
* c       261 Hz    3830    1915
* d       294 Hz    3400    1700
* e       329 Hz    3038    1519
* f       349 Hz    2864    1432
* g       392 Hz    2550    1275
* a       440 Hz    2272    1136
* b       493 Hz    2028    1014
* C       523 Hz    1912     956
```

Figure 39: Note table

Chapter 5: Some Simple Projects

We will keep this as simple as possible and generate these notes using the Arduino library delayMicroseconds() function. (A microsecond is 1/1,000,000 second - yes: one millionth of a second – you have heard that computers are fast haven't you?).

To generate the 'c' note we create an output waveform (see Figure 40: 'c' note waveform) that turns on and off with a frequency of 261 cycles per second. Each of these on/off cycles occurs in 1/261 of a second or 0.003831 seconds. Since we are dealing with microseconds we multiply this by 1,000,000 to get 3831 microseconds per cycle. And since we need to cycle the pin (turn the pin on and off) in that time, we turn it on for 3831/2 = 1915 microseconds (throwing away the fractional part) and off for 1915 microseconds giving us a total of 3830 - we lost 1 due to our not wanting to use fractions, but who is going to miss a microsecond?

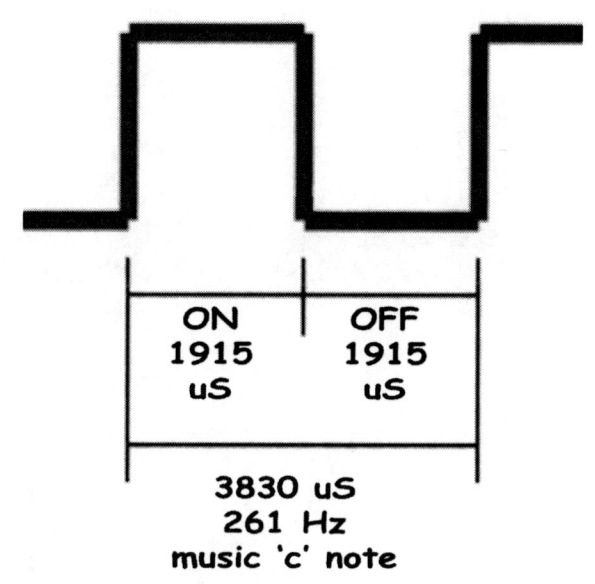

Figure 40: 'c' note waveform

In the Tunes program we use a playTone() function that takes the tone and the duration as parameters. A loop repeats the on/off cycle for the note parameter for a length of time in the duration parameter. It might, for instance, turn the speaker on for 1915 uS (microsecond) and off for 1915 uS repeating for a full second to give a rather long 'c' note. The playTone() function is called by the playNote()

function that has the job of reading through the tune array to get the next note/duration combination.

Each tune is played by an individual function that contains two arrays, one for the tune notes and one for the tune beat. It calculates the duration from the beats and sends that duration along with the tune note to the playNote() function. The playNote() function reads through an array of note names and uses that name position in that array to get the number for the microseconds needed to play that note. It then calls playTone() with the tone microseconds and the duration as parameters. The playTone() uses those parameters to turn the pin connected to the piezo element on and off thus generating exquisite music like none heard since the last pterodactyl blundered into a fern tree.

Happy Birthday Source Code:

```
// This is mostly from the Arduino Melody example:
// (cleft) 2005 D. Cuartielles for K3
// located at:  http://www.arduino.cc/en/Tutorial/Melody
// and modified by Joe Pardue October 1, 2009

int speakerPin = 9;

void setup()
{
  pinMode(speakerPin, OUTPUT);
}

void loop()
{
  play_Happy();
}

void playTone(int tone, int duration) {
  for (long i = 0; i < duration * 1000L; i += tone * 2) {
    digitalWrite(speakerPin, HIGH);
    delayMicroseconds(tone);
    digitalWrite(speakerPin, LOW);
    delayMicroseconds(tone);
  }
}
```

Chapter 5: Some Simple Projects

```
void playNote(char note, int duration) {
  char names[] = { 'c', 'd', 'e', 'f', 'g', 'a', 'b', 'C' };
  int tones[] = {1915, 1700, 1519, 1432, 1275, 1136, 1014, 956 };

  // play the tone corresponding to the note name
  for (int i = 0; i < 8; i++) {
    if (names[i] == note) {
      playTone(tones[i], duration);
    }
  }
}

//Happy Birthday
int Happy_length = 26; // the number of notes
char Happy_notes[] = "ccdcfeccdcgfccCafedbbafgf "; // a space represents a rest
int Happy_beats[] = { 1, 1, 2, 2, 2, 4, 1, 1, 2, 2, 2, 4, 1, 1, 2, 2, 2, 2, 6, 1, 1, 2, 2, 2, 2, 4 };
int Happy_tempo = 150;
void play_Happy()
{
  for (int i = 0; i < Happy_length; i++) {
    if (Happy_notes[i] == ' ') {
      delay(Happy_beats[i] * Happy_tempo); // rest
    } else {
      playNote(Happy_notes[i], Happy_beats[i] * Happy_tempo);
    }

    // pause between notes
    delay(Happy_tempo / 2);
  }
}
```

I was told that there are a couple of sour notes in this. I was surprised that the quality was actually good enough that anyone (Jay - obviously more musical than me) could discern a sour note.

Like with the Cylon Eyes example, we did a simple demonstration using the easier Arduino concepts that will let us play a tune. Later, in Chapter 8: Communicating with a PC, section: 'Jukebox' Tunes Source Code we will make a juke box with multiple noises that can be selected from Arduino Serial Monitor.

71

Chapter 5: Some Simple Projects

Analog

Using PWM to Fade an LED

The Arduino has a built in PWM function: analogWrite() that pulses at about 490 times a second and allows you to set the on time from 0 (off all the time) to 255 (on all the time) or anything in between. We will revisit PWM in more detail in Chapter 10: Simple Motor Speed Control, section: Using PWM to control the motor speed.

We can use the LED attached to pin 9 to demonstrate PWM (Pulse Width Modulation). Figure 41: PWM Fade, shows our board being waved up and down with the LED fading in and out. Notice that the larger streaks (from an LED on the breadboard attached to pin 9) seem like beads whereas the smaller streaks (from the PWR LED on the Arduino board which doesn't blink) are smooth. The visual 'beads' are because the breadboard LED is being turned on and off every ~33.333... times per second (by the delay(30) function). The brightness of those visual beads comes from the setting of the PWM in the analogWrite() function. The camera lens was left open to capture several hundred of those intervals. Look at the center of the large streaks and you will see each 'bead' gets progressively brighter until they seem to blend in the brightest part of the sweep.

The concepts behind PWM are worthy of a full Workshop (or two) but as you will see from the source code, it is actually a fairly simple concept to implement on the Arduino. There are two loops, one for fading in and the other for fading out. Each loop steps through 0 to 255 in increments of 5 and uses that value in the Arduino analogWrite() function that sets length of time the LED is turned on in each cycle. When you look at the LED without shaking the board, it seems to brighten and fade smoothly since the eye/brain smoothes out rapidly blinking lights. The phenomenon is called Persistence Of Vision (POV) and is the same thing that makes movies and TV seem to move smoothly when in fact you are seeing a sequence of still images.

There are no schematics or source code in this example since hooking up an LED on pin 9 was covered earlier and the code to use is the Fading program from the Arduino Example Analog directory.

Chapter 5: Some Simple Projects

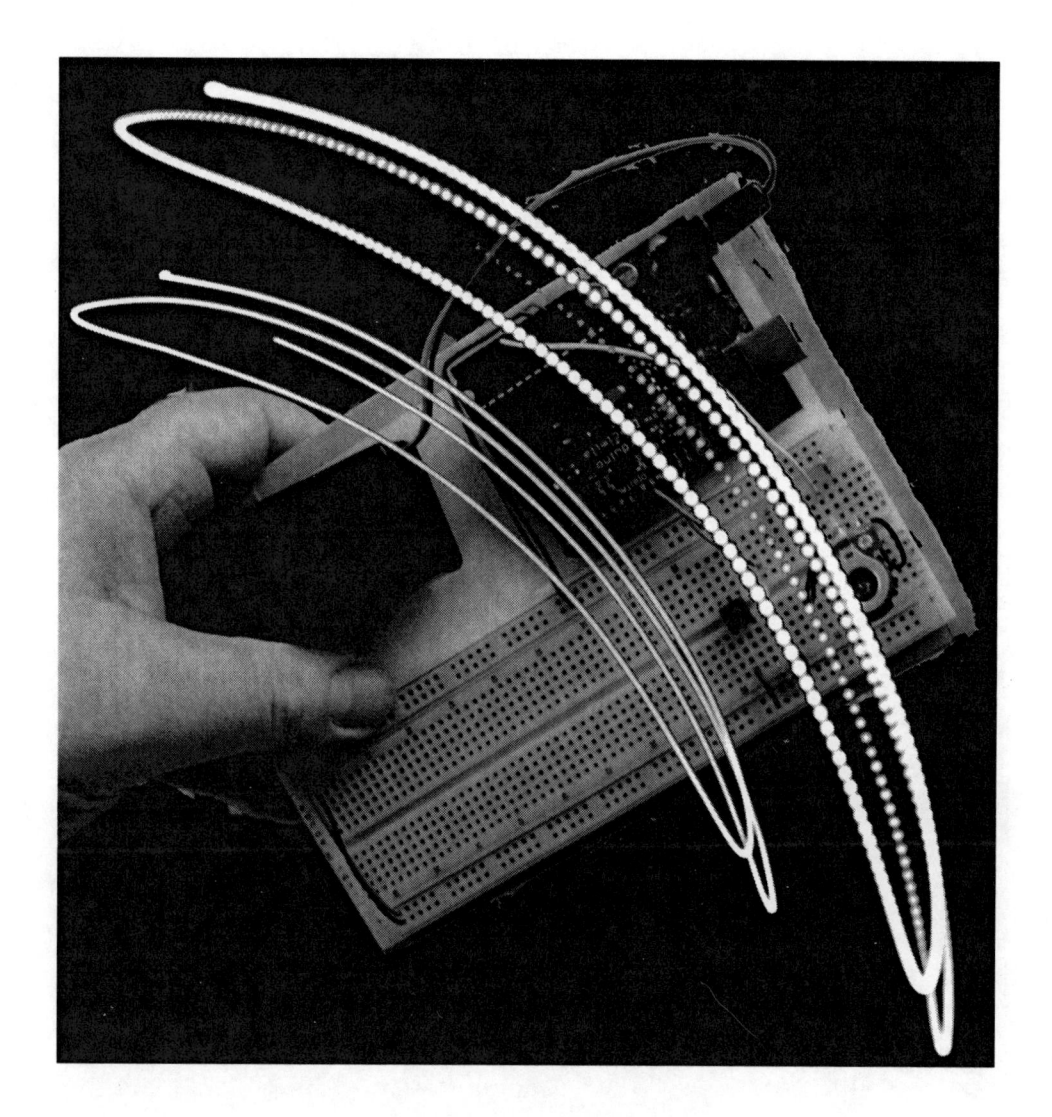

Figure 41: PWM Fade

Chapter 5: Some Simple Projects

Chapter 6: Playing with electricity

If you want to play with electricity, it can help to learn a few facts about the stuff. First fact:

Electricity is dangerous (well, duh!)

Figure 42: Learning the shocking truth about electricity.

Chapter 6: Playing with electricity

When I was about five, my older brother made me angry, so I took a pair of scissors and cut the wire to his record player (which was playing some fool yelling about his 'bucket got a hole in it' - yes it was that long ago) and the next thing I knew, I was up against the opposite wall staring at a melted gap on the scissor blades and glancing over at wall socket which was still smoking and sparking. I will never forget the smell (the burned insulation smelled bad too). I clearly remember the before and after, but the in-between escapes me. I still have those scissors as kind of a reminder that I'm living on a lot of borrowed time. I suspect that the reason I survived was the position of my legs right before I cut the wire. I was scrunched up to the socket with my knees bent and legs to either side thus the shock made my legs straighten out instantly (like in those very creepy frog-leg experiments) and the waxed wood floor provided a nice low friction surface over which my sorry butt could shoot to the opposite wall. I further suspect that my guardian angel was standing next to me slapping out flames on her wings while looking up and begging for a new assignment. [I know that the reality is that I wasn't grounded: if I'd had my other hand on something properly grounded, I wouldn't be writing this.]

We shouldn't have any safety problems since we'll be using small batteries or power from our USB port, but just be aware that if you decide to do anything with that wall plug, you can kill yourself.

Electric Measurements

When we want to bake a cake, we follow a recipe that will say something like: we need two cups of this, a teaspoon of that, as dash of something else, baked at some temperature for some time. If we want to have a cake that is edible, we need to know what the measurements mean. What are cups, teaspoons, dashes, degrees Fahrenheit, and minutes? Likewise to play with electricity we need to know some electrical measurements.

There are three things that we measure when we learn about electricity: Voltage, Current, and Resistance with units in volts, amps, and ohms. Volts are a measure of electric potential (force - how hard it's pushing), amps are a measure of electric current (how much of it is moving through something), and ohms are a measure of opposition to current flow.

Chapter 6: Playing with electricity

Electric Potential Difference = Voltage

Less pressure

More pressure

Same size holes but at different heights - voltage metaphor.

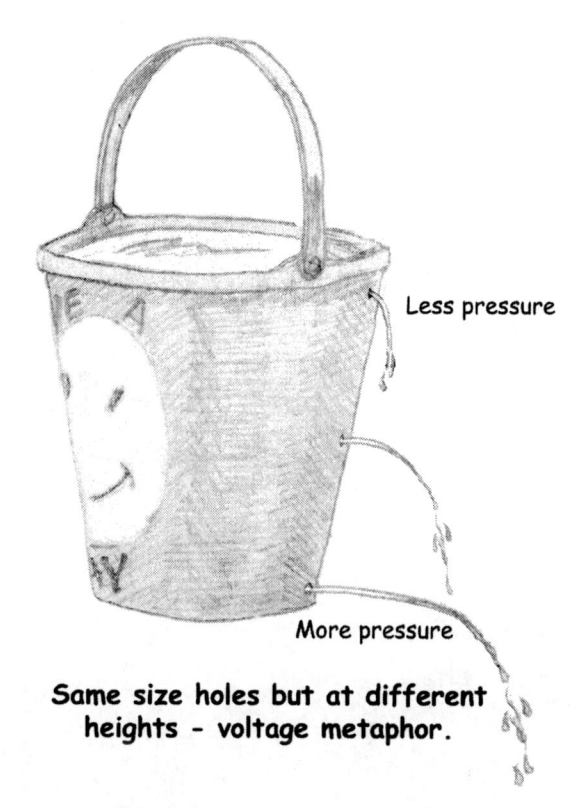

Figure 43: Bucket voltage metaphor

We intuitively understand that if someone dumps a bucket of water off a 20-story building and it hits the sidewalk next to us we are going to get splashed more than if someone standing next to us dumps that bucket of water on the same spot. We use this fact every time we turn on a faucet. The water pipe is connected to a tank of water somewhere that is higher up than our faucet: the higher the tank the greater the water pressure. Voltage is like water pressure, but electric 'pressure' does not come from gravity, it comes from the simple fact that electrons do not like each other and the more crowded they get the madder they get and the more determined they become to bust out and go somewhere with fewer electrons (hey, that's what quantum physicists say if you read between the lines). Areas with more electrons are said to have a higher electric potential relative to areas with

Chapter 6: Playing with electricity

fewer electrons. This potential difference can be thought of as a force that pushes electrons from one place to another.

Voltage is a measure of electric potential difference between two areas that have different amounts of electrons. Water **will** run downhill if it **can** run down hill and electrons **will** spread out if they **can** spread out.

Electric Current = Amps

Less current

Different size holes but at the same height - amp metaphor.

More current

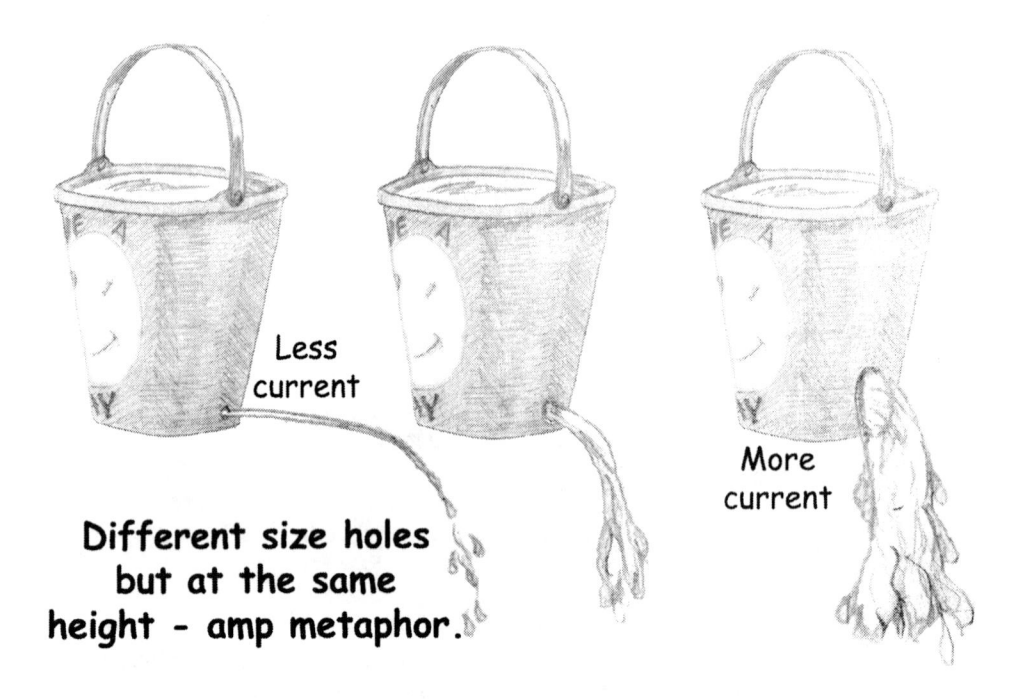

Figure 44: Amp metaphor

We think of current as the amount of water moving past in a stream. The Gulf Stream is a current that has a lot more water moving along than the stream coming out of your bathroom faucet. A bolt of lightning has a lot more electrons moving past than that spark of static electricity your older brother applies to your ear after sliding his bunny slippers across the wool carpet in the hall on a dry day.

The amount of electric current is referred to as Amperes or Amps.

Chapter 6: Playing with electricity

Electric Resistance = Ohms (Ω)

For the water analogy, we can think of resistance as being caused by the diameter of the pipe. Note that in Figure 44: Amp metaphor, more current is due to different size holes – bigger hole, less resistance, more water.

For the resistors in the projects kit, the higher the number the more the resistance so you can think about the 10k Ω resistors as having a smaller electric hole and thus passing less current than the 150 Ω resistors which has a bigger electric hole and passes more current. (The symbol Ω is means Ohm in this context) You can think about the potentiometer from the kit as a kind of variable resistor that works like the valve in a faucet, turn the knob one way the hole gets bigger and the water flow increases, the other and it decreases, we'll learn more about this in a few pages.

Materials with low resistance are known as conductors, while material with high resistance are known as insulators. Electrons move easily in some things such as copper wire but are stopped cold by some things such as glass. Copper has very low resistance – the jumper wires in the projects kit have near 0 ohms resistance (about 1 ohm per 62 feet), while glass can have millions of ohms resistance.

Ohm's Law

A guy named Ohm wrote a rule to account for observations of voltage (V), current (I), and resistance (R). We typically see this rule, as voltage is equal to current times resistance:

$V = IR$ Voltage equals current times resistance.

A little algebra shows us that the equivalents are:

$I = V/R$ Current equals voltage divided by resistance.
$R = V/I$ Resistance equals voltage divided by current.

We will look with this rule (law) in few pages after we introduce some concepts that will help us play with the hardware.

Chapter 6: Playing with electricity

Circuits

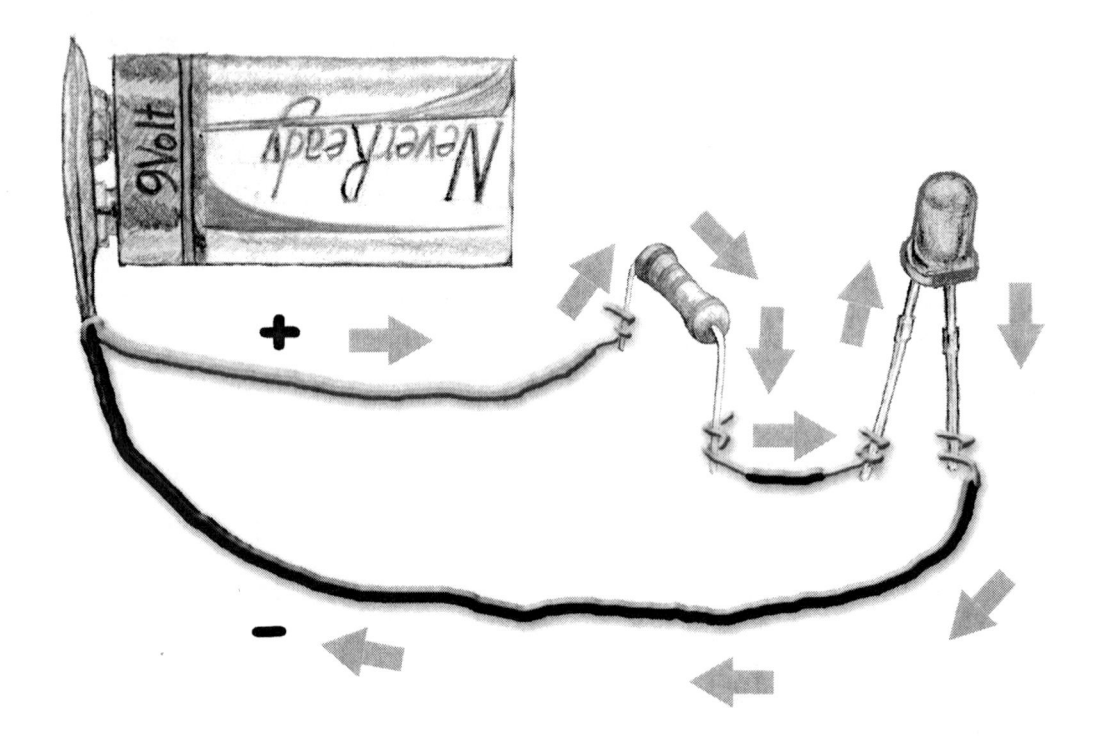

Figure 45: Flow of conventional current

We get electricity to do useful work by channeling it from devices that produce electric force (like generators and batteries) through devices that do electric work (like lights and motors) and back to the device that created the force. That last part is critical. Circuit is just a fancy way of saying 'circle': electricity must run around a circle to do useful work.

Figure 45: Flow of conventional current, shows arrows marking the direction of conventional current from the higher voltage side of a 9-volt battery (the positive terminal) through a resistor and an LED back around to the lower voltage terminal of the battery.

You have probably seen really complex circuits on printed circuit board or as schematics, but no matter how complex it looks, it can be simplified to one part

Chapter 6: Playing with electricity

producing the force as a current, one part using that force to do work, and the circular electrical connection between them.

In case you wonder why we have to add 'conventional' to current, well that is because the guy whose face graces the US 100 dollar bill, yes Benjamin Franklin himself, guessed wrong when he said that charge is carried by positive particles that flow from the + side of a circuit to the – side. It was much later that folks came to realize that the actual charge carrier is an electron, which has a negative charge, and of course, by then the text books all had current going from + to -, so we are still stuck with this backwards concept. And while this won't matter much in anything we'll do here, it caused me no end of confusion in one point of my education, which I won't bore you with, but just be aware if you start trying to understand the physics of transistors, you really need to keep track of which kind of current they are talking about, conventional or electron.

Short Circuits

If we connect a copper wire between the + and – terminals of a battery, 'short circuiting' them, the current will rush through doing a little bit of work making the wire heat up and quickly deplete the chemicals in the battery that are creating the electric potential difference in the first place. Don't try this experiment because not only will it deplete your battery, many batteries will heat up and possibly even explode when treated this way.

If you are doing Arduino experiments plugged into the USB port of your computer and you short circuit the + to the -, and if you are lucky, the USB protection circuits on the PC will detect the current rush and shut down your USB connection before something blows up in your PC. And if you aren't lucky? Well, say bye-bye to something expensive. The morale? Be careful not to short circuit anything expensive, flammable, or with tendencies to explode, that being most things that can be short-circuited – including you.

Chapter 6: Playing with electricity

Electrifying Experiments

An Arduino Volt Meter

Theory is nice, but I'm still a bit like that kid with the melting scissors and I want to get my hands on things and see things happen. To help with this urge, I wrote a PC program that talks to an Arduino and displays the voltage read from the Arduino Analog Input pin 0. We will look at how this code works on the Arduino in a later section. You could easily do these experiments with a hand-held voltmeter, but making your own virtual voltmeter using an Arduino and a PC with the source code for both sides available - how cool is that?

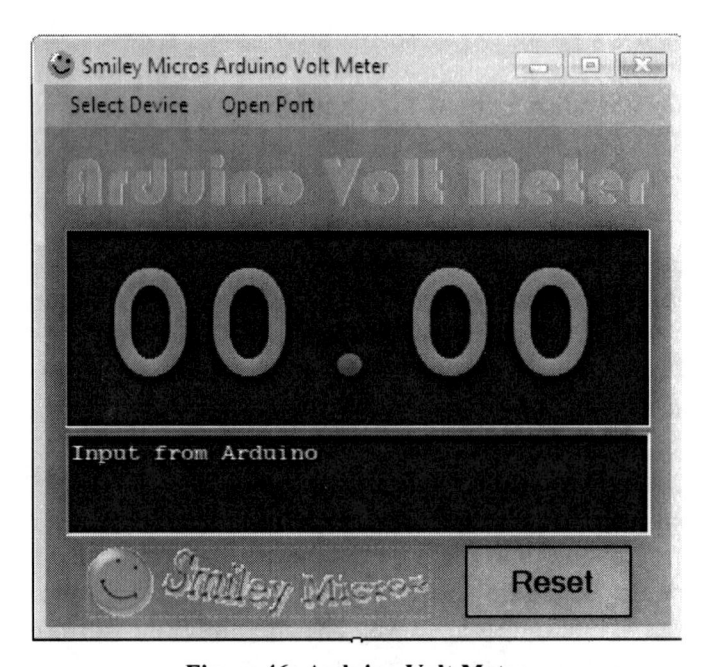

Figure 46: Arduino Volt Meter

The Arduino Volt Meter is written in C# (pronounced C-sharp) Express .NET (free from Microsoft) and the source is available on www.smileymicros.com. I discussed how to write programs like this in some Nuts&Volts articles. If you really are interested in the communications link between the PC and an Arduino, you could do worse than get the book *Virtual Serial Port Cookbook*, but be forewarned, it isn't really for novices.

Chapter 6: Playing with electricity

Installing the Arduino Volt Meter on your PC.

To use the Arduino Volt Meter you need to install The Arduino Volt Meter application that you can download from my website. Just open the directory and click on setup.exe. This program runs on .NET and the installer may demand some Internet time if it isn't already on your system. After installing this program, you will then need to upload the AVM_Test source code (next page) to the Arduino. The Arduino sends an ADC (Analog to Digital Conversion – changing a voltage to a binary number) value once per second to the PC that displays the voltage like it might look on a regular voltmeter and it also displays the raw ADC value in a text window that resembles the Arduino IDE Serial Monitor. Or you can forget the fancy Arduino Volt Meter application and just run the Arduino IDE Serial Monitor which will shown you the ADC value once per second, and you can do the math to convert it to voltage.

Opening the Arduino in the Volt Meter

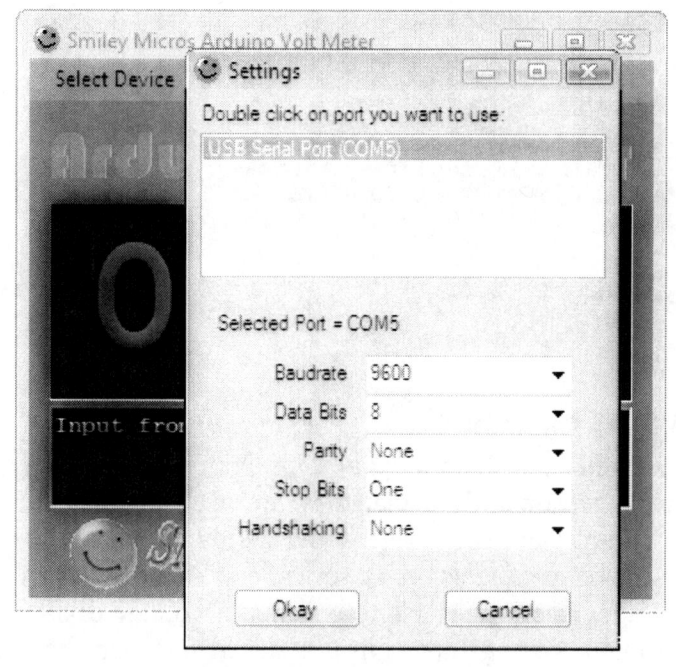

Figure 47: Select your Arduino

Click on the COMx connected to your Arduino and make sure that it shows up in the "Selected Port = COMx" line.

83

Chapter 6: Playing with electricity

AVM_Test Source Code:

```
// AVM_Test
// Joe Pardue December 17, 2009
// based on Tom Igoe's example in the
// Arduino Serial.print(data) documentation

void setup()
{
  // begin the serial communication
  Serial.begin(9600);
}

// variable to hold the analog input value
int analogValue = 0;

void loop()
{
  // read the analog input on pin 0
  analogValue = analogRead(0);

  // print as an ASCII-encoded decimal
  Serial.print(analogValue, DEC);

  // print a terminal newline character so the AVR Voltmeter
  // will know that it has received the full string
  Serial.print('\n');

  // delay 1 second before the next reading:
  delay(1000);
}
```

Voltage across resistance

Let's build a circuit that let's us play with Ohm's Law. We put 10 of the 1K Ω resistors on the breadboard so that they are each connected in series and then connect one end of that series to +5V and the other end to the GND as shown in Figure 48: Schematic of resistors in series, and Figure 49: Layout of resistors in series. This arrangement of resistors is called a voltage divider. The illustrations show the Arduino Analog Input pin 0 attached between the seventh and eighth resistors counting up from the +0V.

Chapter 6: Playing with electricity

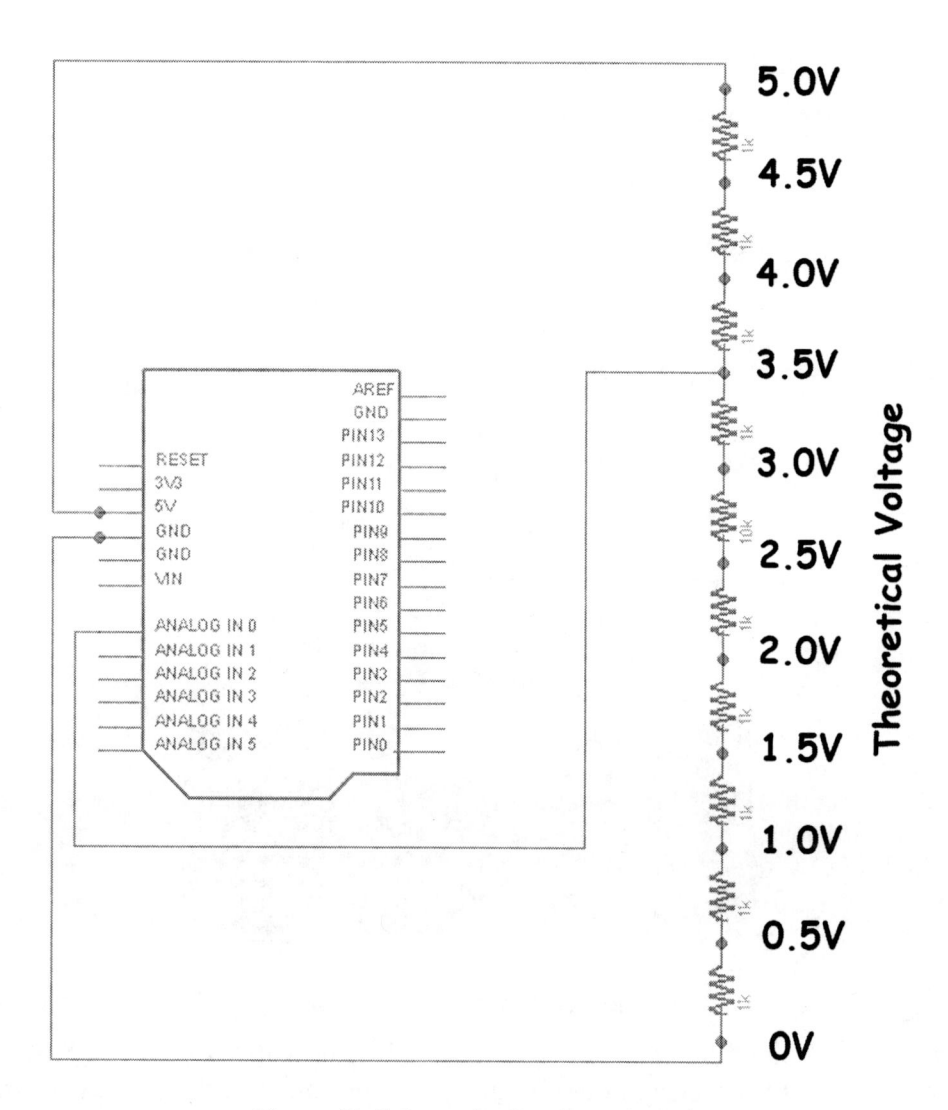

Figure 48: Schematic of resistors in series

Chapter 6: Playing with electricity

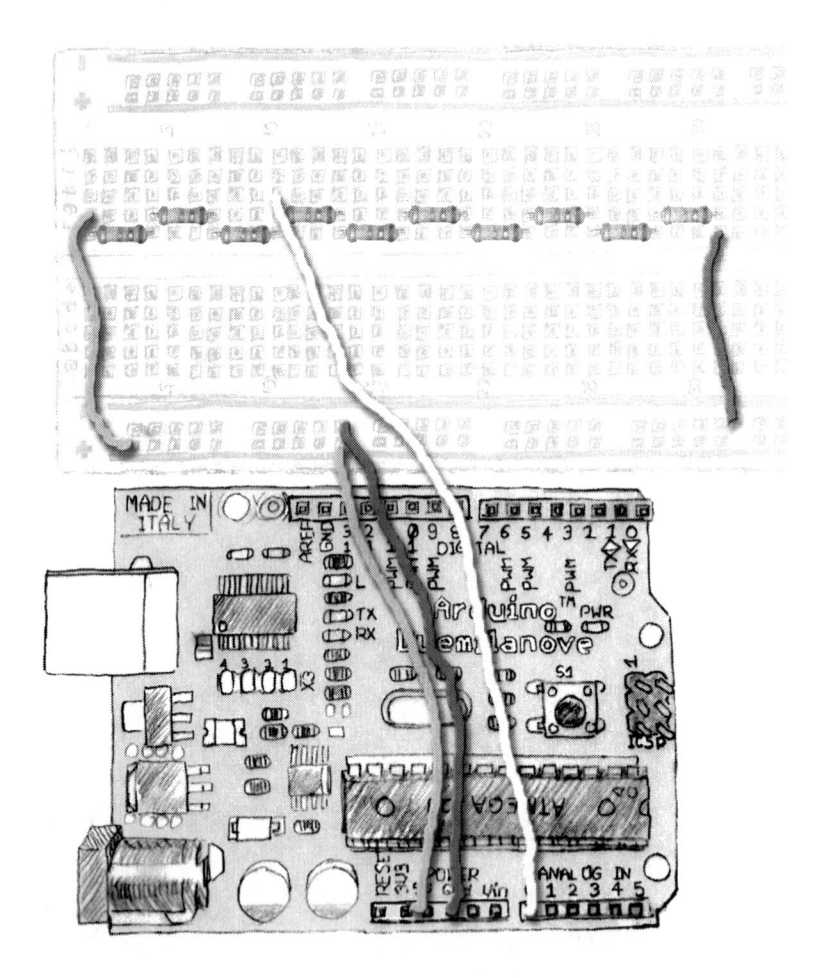

Figure 49: Layout of resistors in series

Let's play with Ohm's Law for a moment. We know that we have 5 volts and a total of 10k Ω resistance in our circuit, so we can calculate the unknown variable, current (I):

I = V/R
I = 5/10000 = 0.0005 Amps

0.0005 Amps is the same as 0.5 milliAmps which we will usually show as 0.5mA. So we have 0.5mA current running through each of the 10 resistors and since each resistor is 1k Ω and we can solve Ohm's Law for the voltage across each resistor:

86

Chapter 6: Playing with electricity

V = IR
V = 0.0005 * 1000 = 0.5 Volts

So, theoretically, we should be able to measure the voltage between the seventh and eighth resistor above 0V (as shown in Figure 48: Schematic of resistors in series) and it should conform to Ohm's Law where the total resistance of the 7 resistors is 7k Ω:

V = IR
V = 0.0005 * 7000 = 3.5 Volts

So let's run the Arduino Volt Meter as shown in Figure 50: AVM measuring resistor divider, and we see we got 3.47 volts.

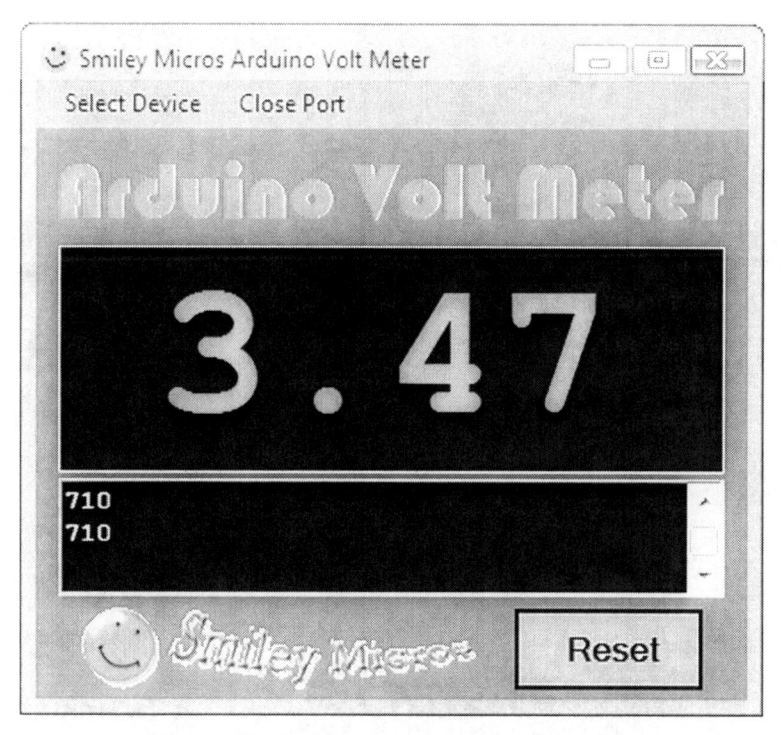

Figure 50: AVM measuring resistor divider

Chapter 6: Playing with electricity

Okay, 3.47 isn't equal to 3.5, but it is about as close as we can expect with a set up like this. Our resistors only claim 5% accuracy so each one can be plus or minus 50 Ω and, again using Ohm's Law we see:

R = V/I
R = 3.47/0.0005 = 6940 Ω

And 6940 is way less than 5% off, and is actually pretty darn close since 5% of 7000 Ω is 350 Ω and we are only 90 Ω off.

Variable Resistance: the Potentiometer

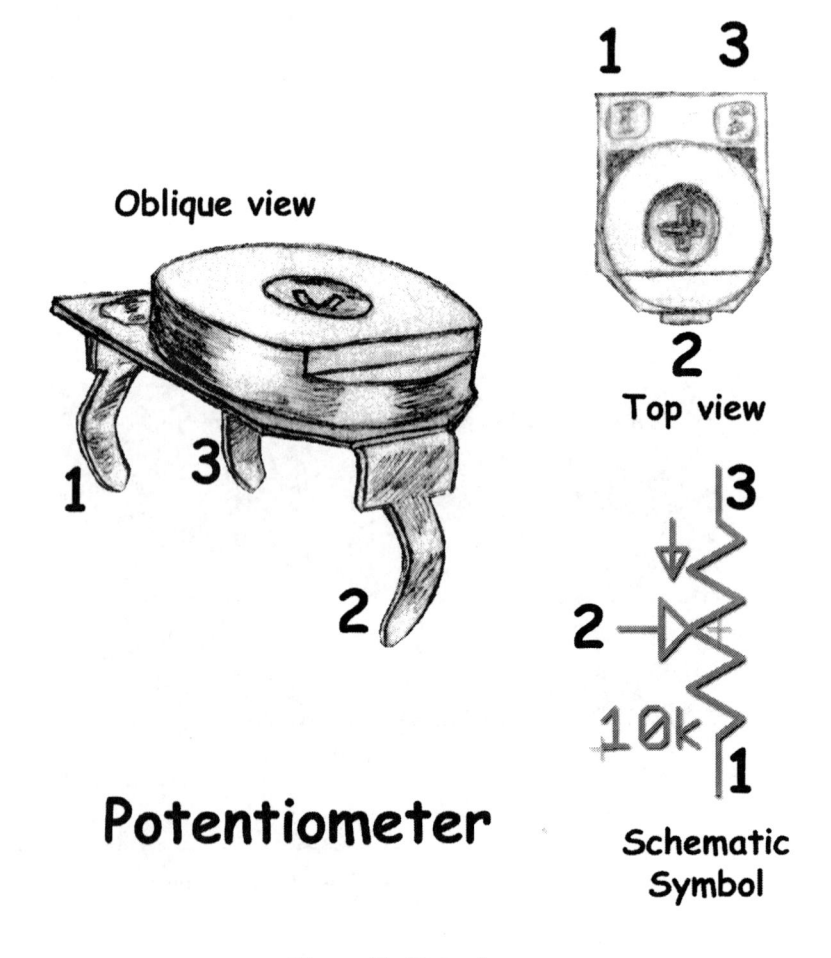

Figure 51: Potentiometer

Chapter 6: Playing with electricity

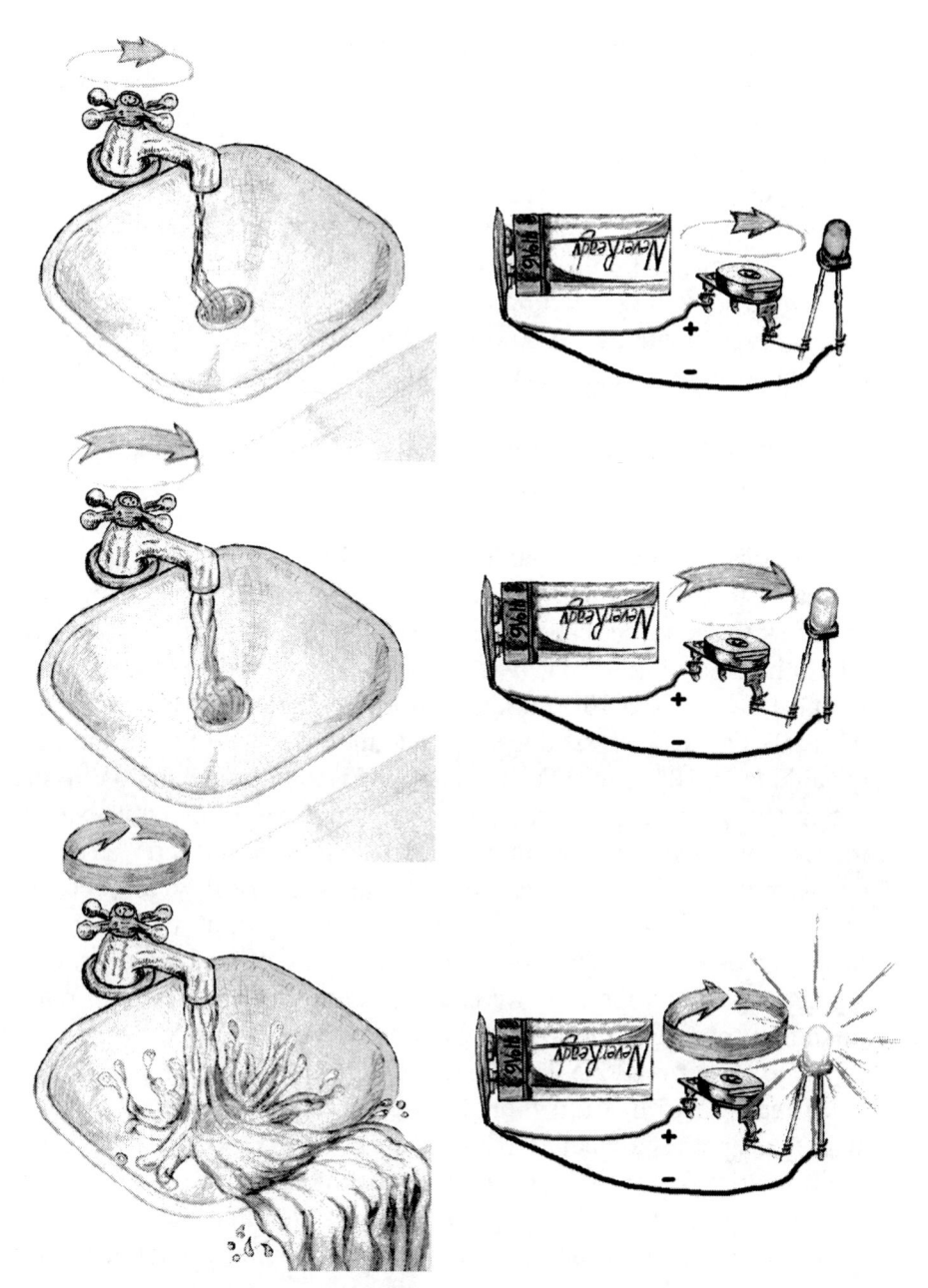

Figure 52: Potentiometer metaphor and real circuit.

Chapter 6: Playing with electricity

A potentiometer is a mechanical device that contains a strip of resistive material connected on each end to a pin, and a third pin in the middle, a wiper, that can be made to slide across the resistive material such that the resistance varies depending on the wiper's position. Figure 51: Potentiometer, shows this concept in the schematic symbol. Pin 2 is the wiper, pins 1 and 3 are on either end. The drawing in Figure 52: Potentiometer metaphor and real circuit., shows that a pot works like a faucet (by the way, don't try the 'real' LED example as shown since a pot can be set to 0 Ω meaning that the full current is applied to the LED and like that sink overflowing, the LED will probably burn up.) [You also don't want to try the faucet experiment since you'll probably look like you peed yourself.]

They are usually called a pot because folks get really tired of typing 'entiometer' every time they want to talk about them (well I do anyway).

A pot works similarly to the resistor divider we saw earlier, except that the connection to the intermediate point is not in discrete steps (in our example 10 steps of 1k Ω each), but can move continuously along the resistance from one side to the other. See Figure 53: The potentiometer wiper, for an illustration of this concept. If we have a 10k Ω pot, then the total resistance between the two outer pins is 10k Ω, but the resistance between the wiper and one of the end pins can be any value from 10k Ω to 0 Ω. If we apply 5V to one side and 0V to the other, then position the wiper in the center we will see 2.5V. Well, not exactly since, like with resistors there are inaccuracies. In the case the pot those errors come from a lack of precision in the resistive material, the inability to precisely move the wiper, and mechanical degradation from sliding that wiper back and forth a bunch. But, as we will see in our experiment, one of the advantages of a pot is that we **can** move that wiper. So we can set the position such that it reads 2.5V even though that setting might not be exactly centered. And if we wear it down a bit so that we no longer read 2.5V in the original position - so what? We just move it till it does read 2.5V. In truth though, cheap pots aren't meant to be moved a lot. Usually cheap ones are called trimmer pots and are moved to a certain setting in a circuit, to trim (compensate) for the inaccuracies in other devices in the circuit, and then they have a dollop of paint or glue poured on them so that they can't be moved. Since our cheap pot is meant for learning purposes it doesn't really matter that it is crappy. If you want a pot that is both accurate and repeatable with lots of use, you can buy them, but they can cost a lot more than you might think.

Chapter 6: Playing with electricity

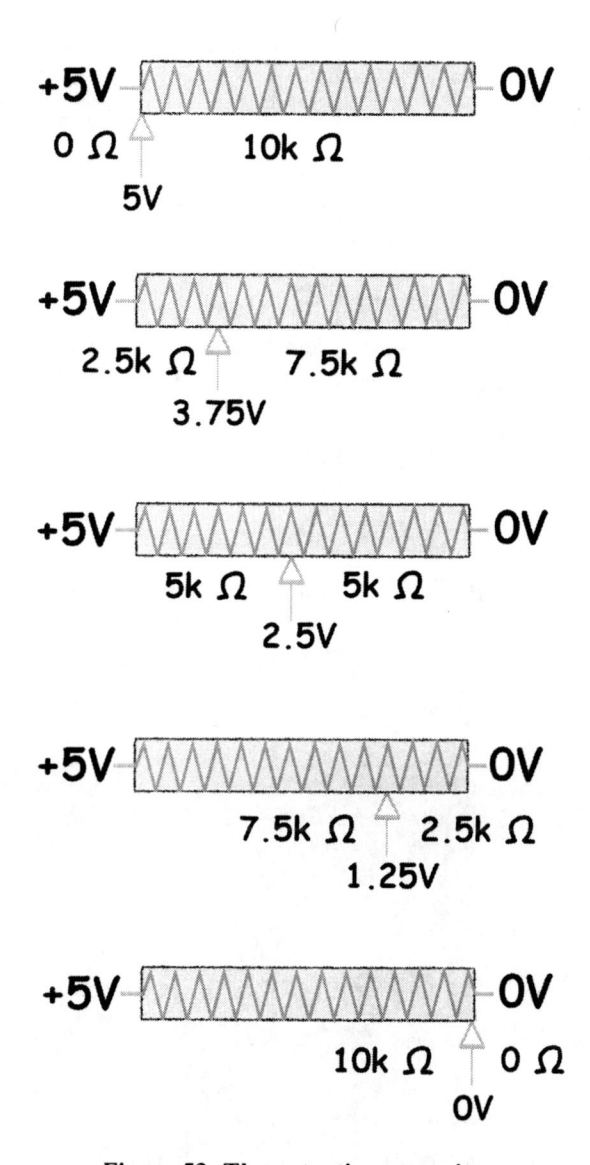

Figure 53: The potentiometer wiper

We can experiment with this by putting our projects kit pot on a breadboard and then wiring it up as shown in Figure 54: Potentiometer to Arduino schematic and Figure 55 Arduino reads potentiometer. You can run the Arduino Volt Meter as before and you'll see all the voltages from 0V to 5V.

91

Chapter 6: Playing with electricity

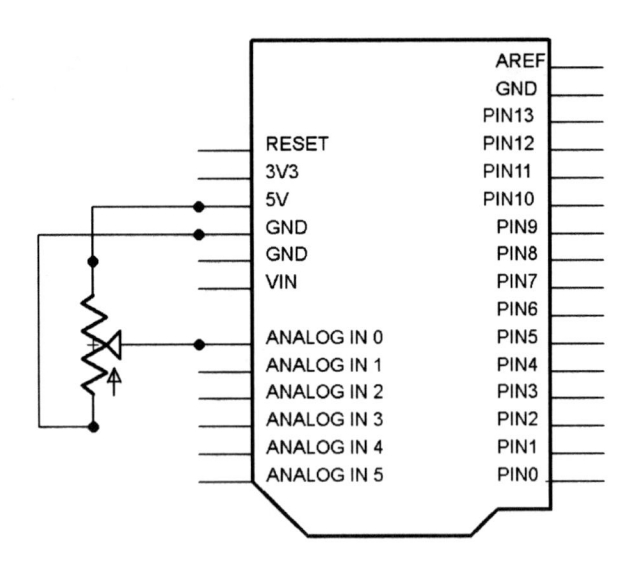

Figure 54: Potentiometer to Arduino schematic

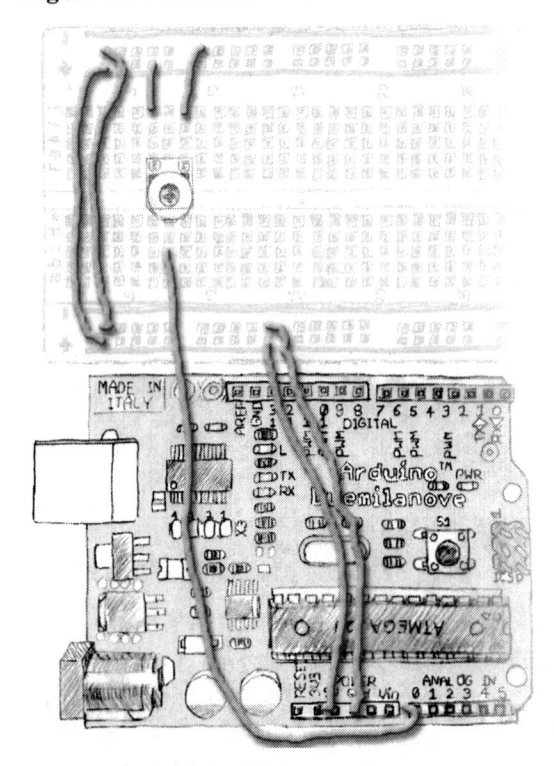

Figure 55 Arduino reads potentiometer

Chapter 6: Playing with electricity

LED Dimmer

We looked at wiring up an LED earlier, but now let's look at using our pot to control the LED brightness and while we are at it, let's measure some voltages and calculate some currents that correlate to the LED brightness.

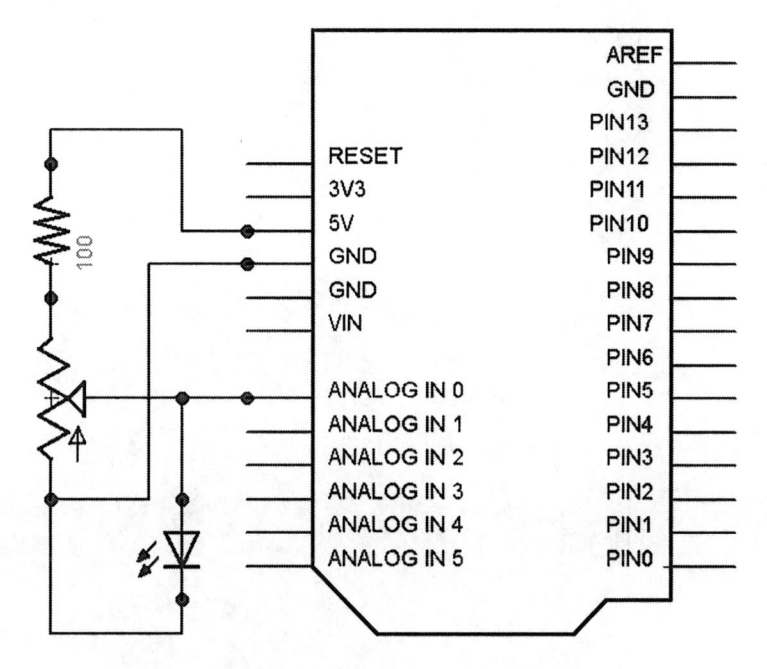

Figure 56: Pot, LED, and Arduino schematic

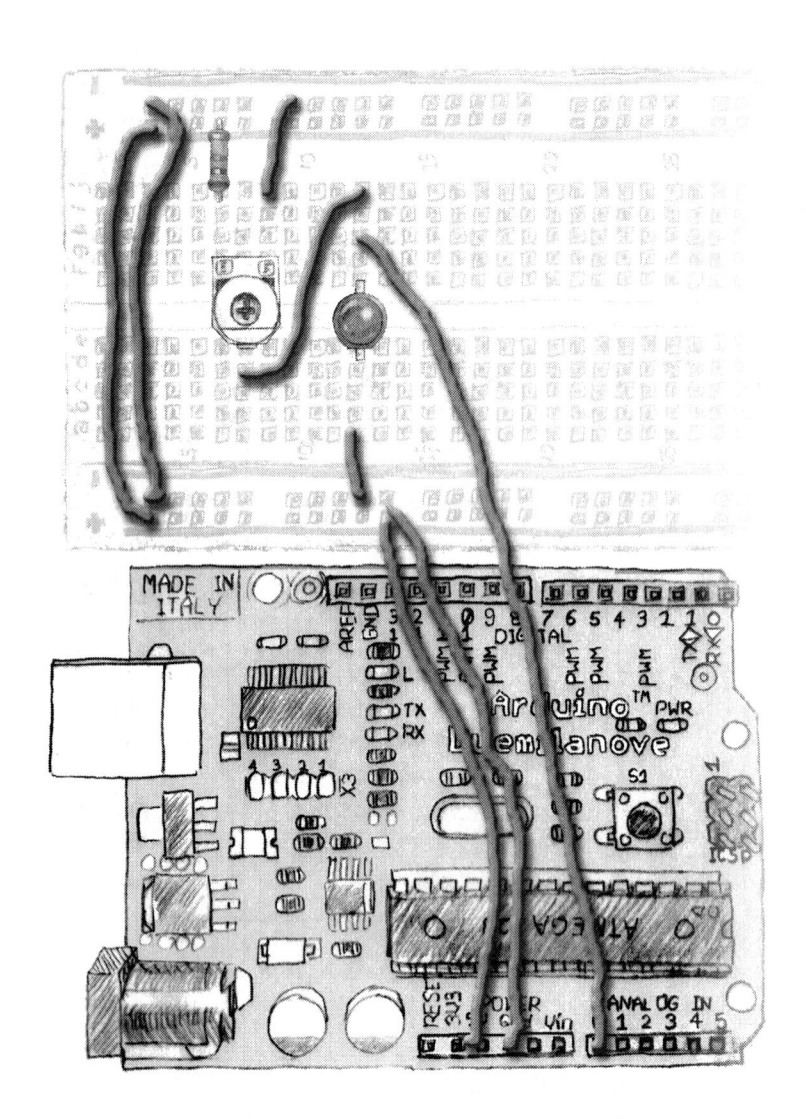

Chapter 6: Playing with electricity

Figure 57: Pot, LED, and Arduino drawing

Follow the schematic and drawings in Figure 56: Pot, LED, and Arduino schematic and Figure 57: Pot, LED, and Arduino drawing, to build the hardware for this experiment. **Carefully note** that we have added a 100 Ω between the pot high volt side and the +5V bus on the breadboard. This resistor is critical because if you set the pot to 0 Ω, **YOU MAY BURN UP YOUR LED.** After you get it wired up and apply the power you can control the LED brightness by rotating the

Chapter 6: Playing with electricity

pot top as before. We know that we are varying the resistance between +5V and the LED + pin (the anode) so that we must be varying the current through the LED to the – pin (the cathode) and that must be what is controlling the brightness.

When you run the Arduino Volt Meter, you will note while the brightness varies from nothing to pretty bright, the voltage varies from about 1.4 to 1.8 volts. At some point when the pot resistance value gets too high, the voltage rapidly drops, but you can't see any light at this point. What we are seeing is that the LED brightness is being controlled by the amount of current passing through it, but within limits.

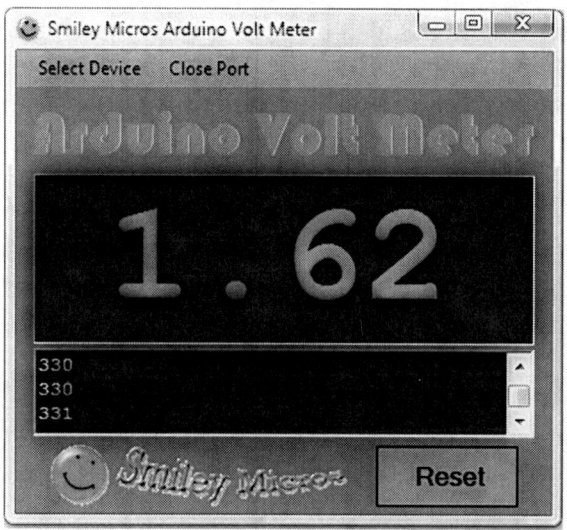

Figure 58: AVM AT 1.62

So let's take some measurements to see if we can understand those limits. I measured my LED dimmest at 1.41V, brightest at 1.83 volts. Note that I'm talking about 'my' LED - yours may differ. So according to Ohm:

    LED at dimmest 1.6V, so 5V - 1.6V = 3.4V
    Current for dimmest I = 3.4V/10100Ω = 0.37mA

[Remember that the pot resistance is 10k Ω but we have an extra 100 Ω so the total is 10100 Ω.]

95

Chapter 6: Playing with electricity

Next I turn the pot till the LED is brightest. In that case the pot is 0 Ω and the 100 Ω resistor is passing all the current:

    LED at dimmest 1.83V, so 5V − 1.83v = 3.17V
    Current for brightest I = 3.17V/100 Ω = 31.7mA

We see that with about 0.2V change in the LED voltage, the current varies from about 0.3mA to over 30mA. We see than for an LED the voltage varies little with the light brightness, but that the current varies a lot. LEDs are rated for an optimum current to maximum light emitting efficiency. They can handle more current for higher brightness at the expense of generating heat that at some point will melt something in the LED. The LEDs in the projects kit were probably swept up off a floor somewhere so no telling what they are rated at, but I notice that using a 1k Ω resistor that drops about 3.3V of the 5V, provides them with:

    Current for my usual setup I = 3.3V/1000 Ω = 3.3mA

Which is probably way less than they can take, but they are plenty bright enough for my purposes so why waste the power?

Okay, we got a little wild there toward the end, but this covers some of the basics concepts useful when playing with electricity. We will look at some more electrifying concepts as we play with the software.

Chapter 7: Playing with software – Part 2

In Chapter 3: Playing with software – Part 1, we learned some introductory concepts and in this section we are going to go a lot deeper with the programming concepts.

Some folks may find this and the following chapters to be very difficult. If you find yourself bogging down I suggest you just skim over the complex stuff and only return to it later if you really want to learn the details. One of the real values of Arduino, as I discussed in Chapter 1, is that Arduino projects can be done without fully understanding the details.

Microcontroller I/O Ports

The first thing is not to confuse the serial communication **ports**, whether USB or COM with the microcontroller I/O **ports**. The former are, as they say, 'serial' meaning that the data comes in a bit at a time in a sequence over one pin (sometimes with some other pins used to control the communication). Parallel port I/O (Input/Output), however, means that the data is presented all at once on several pins, in our case 8, at the same time. So to send the data byte for 10101010 (or hexadecimal 0xAA or decimal 170 which are all the same value but in different number systems) on a serial port, one pin would transition 8 times to send the data. But on a parallel port 8 pins would transition all at once showing the bits side by side. Each way of communicating data has advantages and disadvantages. Serial uses fewer pins and pins are expensive, but parallel can communicate quicker (if you can toggle a serial pin at a certain rate, then sending the same byte of data over a byte-wide port should theoretically be 8 times faster). So you trade-off the number of pins used versus speed. If you are hooking a microcontroller to a PC over 6 feet of wire, then given the cost of copper, serial usually makes more sense, but if you are reading a memory chip located next to a micro on a PCB, then parallel might make more sense.

Let's take a brief look at how the Arduino names the ATmega168/328 pins.

Chapter 7: Playing with software – Part 2

ATmega168/328 Pin Mapping

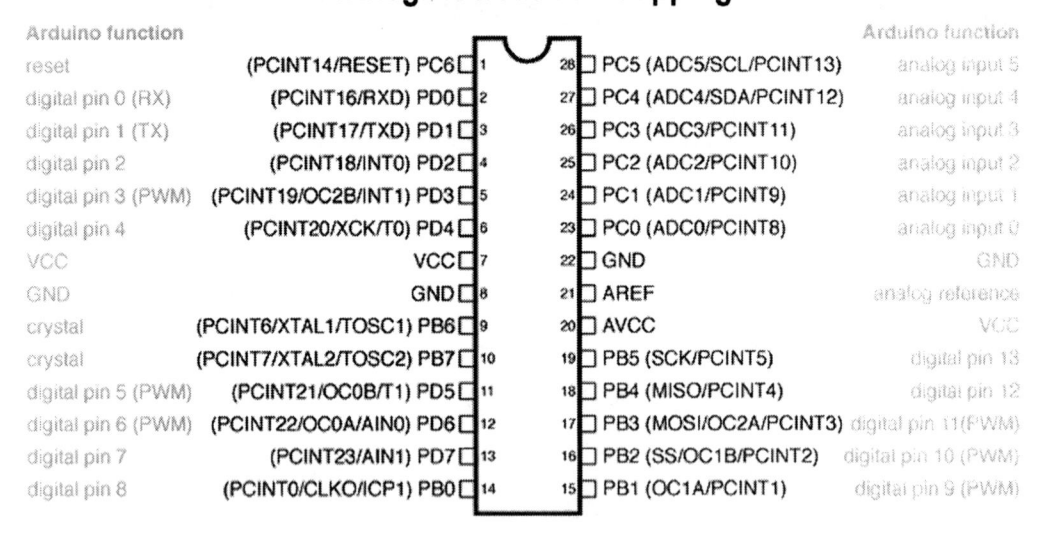

Arduino function				Arduino function
reset	(PCINT14/RESET) PC6	1 28	PC5 (ADC5/SCL/PCINT13)	analog input 5
digital pin 0 (RX)	(PCINT16/RXD) PD0	2 27	PC4 (ADC4/SDA/PCINT12)	analog input 4
digital pin 1 (TX)	(PCINT17/TXD) PD1	3 26	PC3 (ADC3/PCINT11)	analog input 3
digital pin 2	(PCINT18/INT0) PD2	4 25	PC2 (ADC2/PCINT10)	analog input 2
digital pin 3 (PWM)	(PCINT19/OC2B/INT1) PD3	5 24	PC1 (ADC1/PCINT9)	analog input 1
digital pin 4	(PCINT20/XCK/T0) PD4	6 23	PC0 (ADC0/PCINT8)	analog input 0
VCC	VCC	7 22	GND	GND
GND	GND	8 21	AREF	analog reference
crystal	(PCINT6/XTAL1/TOSC1) PB6	9 20	AVCC	VCC
crystal	(PCINT7/XTAL2/TOSC2) PB7	10 19	PB5 (SCK/PCINT5)	digital pin 13
digital pin 5 (PWM)	(PCINT21/OC0B/T1) PD5	11 18	PB4 (MISO/PCINT4)	digital pin 12
digital pin 6 (PWM)	(PCINT22/OC0A/AIN0) PD6	12 17	PB3 (MOSI/OC2A/PCINT3)	digital pin 11(PWM)
digital pin 7	(PCINT23/AIN1) PD7	13 16	PB2 (SS/OC1B/PCINT2)	digital pin 10 (PWM)
digital pin 8	(PCINT0/CLKO/ICP1) PB0	14 15	PB1 (OC1A/PCINT1)	digital pin 9 (PWM)

Digital Pins 11,12 & 13 are used by the ICSP header for MISO, MOSI, SCK connections (Atmega168 pins 17,18 & 19). Avoid low-impedance loads on these pins when using the ICSP header.

Figure 59: Mapping ATmega168/328 to Arduino pins

I/O pin naming is one of the things that Arduino does that is a bit different from what you'll usually see when you learn about micros. It considers the 14 Digital Input/Output pins as individuals rather than one of eight pins in a port. As shown in Figure 59: Mapping ATmega168/328 to Arduino pins. The ATmega168/328 has four ports: B, C, and D labeled in sequences like PB0…PB7 for PortB0 to PortB7. Port C only has 7 pins. And not all pins are actually available, for instance in the Arduino hardware, PC6 is used for resetting the microcontroller. Figure 60: Arduino Pin9 is ATmega328 PortB Pin1, shows how the Arduino maps a port pin to a digital I/O pin.

Chapter 7: Playing with software – Part 2

Figure 60: Arduino Pin9 is ATmega328 PortB Pin1

In Chapter 5 we used the Arduino pinMode(), digitalRead() and digitalWrite() functions to individually read or write to pins in the DIP to LED program. We will now repeat that functionality, but use ports instead of pins.

DIPLED_With_Ports Source Code:

```
// DIPLED_With_Ports
// Joe Pardue December 17, 2009

void setup()
{
 // Init port pin
 DDRB = 0x00; // set port B for input
 PORTB = 0xFF; // set port B pullups
 DDRD = 0xFF; // set port D for output
}

void loop()
{
   PORTD = PINB; // Show the port B input on Port D output
}
```

99

Chapter 7: Playing with software – Part 2

You see that in this version we use a single line in the loop() function to accomplish what it took 5 lines in the original version shown below:

```
void loop()
{
  digitalWrite(0,digitalRead(8));  // show DIP 0 state on LED 0
  digitalWrite(1,digitalRead(9));  // show DIP 1 state on LED 1
  digitalWrite(2,digitalRead(10)); // show DIP 2 state on LED 2
  digitalWrite(3,digitalRead(11)); // show DIP 3 state on LED 3
  digitalWrite(4,digitalRead(12)); // show DIP 4 state on LED 4
}
```

Later in this chapter we will write a much more capable version of the Cylon Eyes program from Chapter 5 and you can how extremely long the program would be if you used only the Arduino pinMode(), digitalRead(), and digitalWrite() functions versus using ports. This clearly shows why lumping pins into ports is such a big deal on microcontrollers.

BTW we used the following lines that we don't quite understand at this point.

```
  DDRB  = 0x00; // set port B for input
  PORTB = 0xFF; // set port B pullups
  DDRD  = 0xFF; // set port D for output
```

But for the time being, Let's just copy and paste the port setup statements and worry about what they mean later.

Redo Cylon Eyes using ports

We can see further advantages for using ports by redoing the Cylon Eyes 1 program from Chapter 5. This version is much shorter.

Cylon Eyes 2 Source Code:
```
// Cylon_Eyes-2
// Joe Pardue October 2, 2009

void setup()
{
  // set port D for output
```

Chapter 7: Playing with software – Part 2

```
  DDRD = 0xFF;
}

void loop()
{
  cylonEyes();
}

/*
Binary and hexadecimal equivalent
00000001 == 0x01
00000010 == 0x02
00000100 == 0x04
00001000 == 0x08
00010000 == 0x10
00100000 == 0x20
01000000 == 0x40
10000000 == 0x80
01000000 == 0x40
00100000 == 0x20
00010000 == 0x10
00001000 == 0x08
00000100 == 0x04
00000010 == 0x02
*/
void cylonEyes()
{
  uint8_t i = 0; // uint8_t means an unsigned 8-bit number
  // ce[] is an array - a collection of numbers
  uint8_t ce[] = {0x01,0x02,0x04,0x08,0x10,0x20,0x40,0x80};

  while(1)
  {
    for(i = 0; i <= 7; i++)
    {
      PORTD = ~ce[i];// '~' will be defined later
      delay(100);
    }
    for(i = 6; i >= 1; i--)
    {
      PORTD = ~ce[i];
      delay(100);
    }
  }
}
```

Operators

Remember when I said it was going to get hard? Here we go... Operators are symbols that tell the compiler to do things such as set one variable equal to another, like the '=' operator, as in count = 0. Operators are much like what you see in Algebra, but not exactly so be careful. In the Arduino Language Reference (discussed above) you see four operator groups:

- Arithmetic Operators
  - = (assignment operator)
  - + (addition)
  - - (subtraction)
  - * (multiplication)
  - / (division)
  - % (modulo)
- Comparison Operators
  - == (equal to)
  - != (not equal to)
  - < (less than)
  - > (greater than)
  - <= (less than or equal to)
  - >= (greater than or equal to)
- Boolean Operators
  - && (and)
  - || (or)
  - ! (not)
- Compound Operators
  - ++ (increment)
  - -- (decrement)
  - += (compound addition)
  - -= (compound subtraction)
  - *= (compound multiplication)
  - /= (compound division)

In the Light_Switch program in Chapter 3 we use several of these operators to check to see if the button has really been pressed or if we are just reading some noise on the button pin. In the loop() function we repeatedly call the checkButton() function and each time it reads the button pressed it counts up, and

Chapter 7: Playing with software – Part 2

each time it reads the button not pressed we count down. We assume that if the count reaches a certain value, in this case DEBOUNCE, then the button really is being pressed. But lets not get bogged down with the topic of debouncing switches since our goal here it to see some operators in action.

```
// Check the button and count up if it is connected to
// ground (LOW) or count down if it is connected to +5 (HIGH)
void checkButton()
{
  // read the pin HIGH or LOW?
  reading = digitalRead(BUTTONPIN);

  // if it is LOW and the count is less than DEBOUNCE
  // increase the count by 1.
  if ( (reading == LOW) && (count < DEBOUNCE) )
  {
    count++;
  }
  else// otherwise if it is greater than 0, decrease by one
  {
    if(count > 0) count--;
  }
}
```

First we set the reading variable equal to the value returned by the digitalRead(BUTTONPIN) function:

```
reading = digitalRead(BUTTONPIN);
```

Next we then check to see if the reading value is now *equal to* the LOW constant **and** if the count variable is *less than* the DEBOUNCE constant:

```
if ( (reading == LOW) && (count < DEBOUNCE) )
```

If they are, then we *increment* (add one to) the count variable:

```
count++;
```

If one (or both) of the two cases wasn't true then if the count variable is *greater than* 0, we *decrement* (subtract one from) the count variable:

```
if(count > 0) count--;
```

Chapter 7: Playing with software – Part 2

Some equals are more equal than others

Of all the operators the '=' and '==' give most folks fits. Not just newbies but careless experienced folks (like me) mess up with these more often than we would like to admit. '=' is the arithmetic **assignment operato**r and it will assign the value from the right side of the = sign to the variable on the left side. '==' is the **comparison operator** and it is used to compare the values on either side of the operator, if they are actually equal then the comparison is said to be true and if they aren't equal then the comparison is said to be false (giving the operation a value of 0). This will bite you as follows:

```
//BAD CODE:
if(ledState = true)
{
        // do something
}
```

This 'if' evaluation will always be true because you just set the ledState to equal true. You really meant to ask a question 'is ledState equal to true?':

```
//GOOD CODE:
if(ledState == true)
{
        // do something
}
```

The confusion involved between these two rules is just part of the price you pay to be a 'real' programmer. You will screw this up so get ready to smack the heel of your hand into your forehead and say: 'Ah, man he said I'd do this!' [Also, a friend told me to recommend that folks reread the past couple of pages before proceeding.]

There are exactly 10 types of people in the world. Those who understand binary and those who don't.

If that doesn't make sense, it will in a few minutes.

Bits

At one time 'computer' was a job description for people with quill pens who spent their lives calculating tables of things like cannonball trajectories to help soldiers more accurately slaughter their enemies. Later mechanical computers, with brass

Chapter 7: Playing with software – Part 2

gears and cams, were developed to make calculating the slaughter cheaper, quicker, and easier. Then one day a genius figured that you could do all this computing even easier if you used switches. Switches can be off or on, and the fundamental datum is the 'bit' with exactly two 'binary', states. We variously refer to these states as '0 and 1' or 'on and off' or 'clear and set' or 'true and false'. It's the latter that allows us to use bits to automate Boolean Logic (an 'algebra' for determining if a statement is true or false) and thus the modern binary logic computer entered the world and now slaughter is so cheap, quick and easy to compute that anybody can do it. Maybe this is skimming the topic a bit (har!) but a full explanation would begin with the first sentence of Genesis and only hit its stride about the time Alan Turing offed himself as his unjust reward for saving the free world in WWII, and while fascinating, it won't get us to blinking LEDs any quicker, so let's move on.

When we use the Arduino to blink an LED, it is connected to a microcontroller pin that can have two voltage states: ground or Vcc, which can be manipulated as a data bit.

Bytes

The AVR and many other microcontrollers physically handle data in 8-bit units called bytes. A **data type** that can have 256 states, 0 thru 255. This is shown in the following sequence of states, (leaving out 5 thru 250):

```
00000000 = 0
00000001 = 1
00000010 = 2
00000011 = 3
00000100 = 4
...(states 5 thru 250)...
11111011 = 251
11111100 = 252
11111101 = 253
11111110 = 254
11111111 = 255
```

Chapter 7: Playing with software – Part 2

Binary numbers have only two digits: 0 and 1 and are built in magnitude from the right to the left. You can see the full sequence in Appendix 2: Decimal, Hexadecimal, and Binary.

Okay, so what are the 10 types of people again?

If you look at the bits in a byte and think of them as LEDs you can visualize a series of bytes that would generate a Cylon Eyes sweep from right to left. What you are seeing is 8 of the 256 possible states being presented in a sequence that fools us into thinking we are seeing a back and forth scrolling motion. Using binary numbers where the lit LED is represented by 1 shown next to the **binary**, **hexadecimal** and **decimal** equivalent, what we are seeing is:

```
00000001 = 0x01 = 1
00000010 = 0x02 = 2
00000100 = 0x04 = 4
00001000 = 0x08 = 8
00010000 = 0x10 = 16
00100000 = 0x20 = 32
01000000 = 0x40 = 64
10000000 = 0x80 = 128
01000000 = 0x40 = 64
00100000 = 0x20 = 32
00010000 = 0x10 = 16
00001000 = 0x08 = 8
00000100 = 0x04 = 4
00000010 = 0x02 = 2
00000001 = 0x01 = 1
```

In microcontroller applications, we will often be dealing with the states of byte-sized ports, like PORTD. A microcontroller port is a place where outside voltages (0V or 5V [or 3.3V or whatever voltage your micro uses) can be read or set.

Experienced microcontroller programmers memorize the binary equivalent of hexadecimal (hex) digits and find hex numbers very useful. For instance, given 0xA9, what would the LEDs (or the voltage states of an 8-bit register) look like? If you memorize the hex table, you come up with 0xA = 1010 and 0x9 = 1001, so

106

Chapter 7: Playing with software – Part 2

the LEDs (voltage states) will look like: 10101001. If you ask the same question in decimal equivalent of 0xA9: "what will 169 look like on the LEDs?", well good luck on doing that in your head. You don't really need to know this for 'playing with software' but such factoids might help.

Bitwise Operators

Table 1: Bitwise Operators

Operator	Name	Example	Defined
~	Bitwise complement NOT	~x	Changes 1 bits to 0 and 0 bits to 1
&	Bitwise AND	x&y	Bitwise AND of x and y
|	Bitwise OR	x|y	Bitwise OR of x and y
&=	Compound AND	X &= y	Equivalent: x = x&y
|=	Compound AND	X |= y	Equivalent: x = x|y
^	Bitwise exclusive OR	x^y	Bitwise XOR of x and y
<<	Left shift	x<<2	Bits in x shifted left 2 bit positions
>>	Right shift	x>>3	Bits in x shifted right 3 bit positions

Bitwise operators are critically important in microcontroller software. They allow us to do many things that can be directly and efficiently translated into microcontroller machine operations. This is a dense topic, so get out a pencil and piece of paper and work through each of the examples until you understand it. Okay, Simon says: 'get out a pencil and piece of paper and work through each of the examples until you understand it.' Or just breeze over it and refer back if you really need it.

The TRUTH about numbers

In case you've ever wondered how to tell what is true and what is false, well… for **bitwise operators** which use binary logic (single bits), 1 is true and 0 is false. (And often this is defined as 0 is false and anything else it true, so 1 would be true as would 2 or 65535). Logic truth is determined by rules in truth tables.

Look at the truth tables for AND '&', OR '|', XOR '^', and NOT '~':

Chapter 7: Playing with software – Part 2

```
     AND            OR            XOR           NOT
0 & 0 = 0      0 | 0 = 0      0 ^ 0 = 0     ~1 = 0
0 & 1 = 0      0 | 1 = 1      0 ^ 1 = 1     ~0 = 1
1 & 0 = 0      1 | 0 = 1      1 ^ 0 = 1
1 & 1 = 1      1 | 1 = 1      1 ^ 1 = 0
```

These truths, while not self-evident, are the foundation for computer decision-making.

ORing

We can set (make it 1) bit 3 in a variable: myByte, by using the bitwise OR operator: '|'

```
myByte = 0;
myByte =  myByte | 0x08;
```

To see what's happening look at these in binary:

```
bit #      76543210
myByte = 00000000 = 0x00
  0x08 = 00001000 = 0x08
-----------------------
    OR = 00001000 = 0x08
```

We see that bit 3 is 1 in 0x08 and 1 | 0 = 1 so we set bit 3 in myByte.

Suppose myByte = 0xF7:

```
bit #      76543210
myByte = 11110111 = 0xF7
  0x08 = 00001000 = 0x08
-----------------------
    OR = 11111111 = 0xFF
```

Or maybe myByte = 0x55:

```
bit #      76543210
myByte = 01010101 = 0x55
  0x08 = 00001000 = 0x08
-----------------
    OR = 01011101 = 0x5D
```

This shows that only bit # 3 of myByte is changed by the OR operation. It is the only bit equal to 1 in 0x08 and according to the OR truth table, ORing 1 with anything is always yields 1 so you can use it to '**set**' a bit regardless of that bit value.

ANDing

We can clear (make it 0) bit 3 in a variable: myByte, by using the bitwise AND operator: '&'

```
myByte = 0xAA;
myByte =  myByte & 0xF7;

bit #     76543210
myByte = 10101010 = 0xAA
  0xF7 = 11110111 = 0xF7
------------------------
   AND = 10100010 = 0xA2
```

Or maybe myByte = 0x55:

```
bit #     76543210
myByte = 01010101 = 0x55
  0xF7 = 11110111 = 0xF7
------------------------
   AND = 01010101 = 0x55
```

From this you see that ANDing with 1 leaves the bit value the same as the original bit, and ANDing with 0 clears that bit regardless of its state so you can use ANDing with 0 to '**clear**' a bit value.

Let me repeat: to set a bit regardless of the bit value you use '|' (OR), to clear a bit you use '&' (AND). You can think of this with the question: 'how would I turn an LED on or off regardless of whether it is on or off to begin with?' The answer: 'You would '|' with the bit controlling the LED pin to turn it on, and you would '&' that bit with 0 to turn it off.

Chapter 7: Playing with software – Part 2

Setting and Clearing Bits

In each of the above cases we are only dealing with a single bit, but we might be interested in any number of the bits in a byte. Another important feature of using bitwise operators is that it allows us to set or clear a specific bit or group of bits in a byte without knowing the state of or affecting the bits we aren't interested in. For example, suppose we have eight lights whose on/off state is controlled by a byte and we want to turn on light six and turn off lights 0 and 2 – regardless of their current state. Let's set bit 6, regardless of its present value, then clear bits 0 and 2, also regardless of their present value and, here's the trick, we must leave bits 1, 2, 4, 5, and 7 as they were when we began.

NOTE: `myByte =  myByte | 0x40;`

Is the same as `myByte |= 0x40;`

Both read as 'myByte is equal to the results of ORing myByte with 0x40.

To set bit 6 we OR myByte with 01000000:

```
        myByte = 0x2B;
        myByte |= 0x40;

bit #      76543210
myByte = 00101011 = 0x2B
         01000000 = 0x40
         ------------------------
    OR = 01101011 = 0x6B
```

Next we want to clear bits 0 and 2 so we AND myByte with 11111010:

```
        myByte &= 0xFA;

bit #      76543210
myByte = 01101011 = 0x6B
  0xFA = 11111010 = 0xFA
         ------------------------
   AND = 01101010 = 0x6A
```

Chapter 7: Playing with software – Part 2

So in summary we set bits with '|' and clear bits with '&'. We will later look at using these operators with **bit masks** that are patterns of bits that are used for the retention or elimination of another pattern of bits.

In case you are asking 'why on Earth would I ever want to do anything like this?' Think of the case for a dashboard on a car or the front panel of a kitchen appliance, you might have 8 LEDs indicating various things and want to control those LEDs using a microcontroller's 8-bit port. Bitwise operators are how you do that.

XORing

Suppose we want to flip (change 0 to 1 and 1 to 0) the highest 4 bits in a byte while leaving the lowest 4 alone, we could use a mask with the bits you want to flip set to 1 and the bits you don't want to flip set to 0:

```
myByte = 0xAA;
myMask = 0xF0;

myByte ^= myMask; // ^ is the XOR operator

bit #    76543210
myByte = 10101010 = 0xAA
myMask = 11110000 = 0xF0
------------------------
   XOR = 01011010 = 0x5A
```

XORing is used a lot in cryptographers, but not a lot by us mortals.

NOTing

Using the above example, we could clear the mask bits in myByte by using the NOT operator on the mask then AND it with myByte.

```
myByte = 0xAA;
myMask = 0xF0;

 myMask = 11110000 = 0xF0
~myMask = 00001111 = 0x0F // ~ is the NOT operator
```

111

Chapter 7: Playing with software – Part 2

```
   myByte &= ~myMask; // & is the AND operator

   bit #      76543210
   myByte =   10101010 = 0xAA
  ~myMask =   00001111 = 0x0F
            ----------------------
       AND =  00001010 = 0x0A
```

If you are going 'Oh my God!' at this point, I hope it is because you are surprised that you actually understand this. If it isn't, then get that pencil and paper and go back till you get it (err... Simon says...). If you stick with micros this 'stuff' will become useful one day. I promise.

Bitwise versus Boolean Operators

Bitwise operators are used to change values, Boolean operators are meant to evaluate logical statements to see if they are true or false. Folks tend to confuse the bitwise and Boolean operators, especially the & versus && and the | versus ||. This is analogous to the = versus == operators. The operator with the single symbol: =, &, | are each used to change a value in a computation while the operator dual symbol: ==, &&, and || are used to compare values for making a decision.

If we use | (OR) we are performing a computation that changes a value as shown a couple of pages before: `myByte = myByte | 0x08`. The value of myByte may be changed by this operation. If we use the || which is also 'OR ' we are asking the micro to look at two values and see if **either** one of them is true. Likewise for &&, we are asking if **both** values are true. You may have been exposed to this sort of logic at one time with a list of statements like:

   Only birds have beaks and feathers
   Polly has a beak and feathers
   Therefore Polly is a bird

This statement asserts the two truths may be sufficient to prove that something is a bird: beaks && feathers, so we can pose the test:

Chapter 7: Playing with software – Part 2

```
boolean pollyBeak = true;
boolean pollyFeathers = true;

if(pollyBeak && pollyFeathers)
{
     Serial.println("Polly is a bird.");
}
else
{
     Serial.println("Polly may not be a bird.");
}
```

And note that the decision is only that Polly is a bird, but we don't have enough information to decide that Polly is not a bird since we didn't assert that beaks and feathers are a requirement for being a bird. We could have said **all** rather than **only**: 'All birds have beaks and feathers', then the last statement could have been "Polly is not a bird". Logic can get kind of hairy, but since this is not a book about logic (okay it is, but the logic is well buffered) we won't go further with this.

Well, deciding if Polly is a bird or not might seem like a weird thing to go to so much trouble to figure out. But what if you want to design one of those cool two-man key switches that are needed to launch the missiles (and often have folks shouting and pointing guns at each other in the movies)? The && operator is just what you need.

We will reuse the hardware from the Chapter 5 DIP Switch and LEDs project (see Figure 35: DIP switch and LED schematic.)

In the following Launch_Control program, the key point is in the 'if' line:

```
if(switchOneState && switchTwoState)
{
   // do something if both variables are true
}
```

The above if() statement will only run the statements between the following open and close brackets: {} *if* both switch state variables are *true*. The full program follows.

Chapter 7: Playing with software – Part 2

Launch_Control Source Code:

```
// Launch_Control
// Joe Pardue October 3, 2009
// If you use the DIP LED setup for this then LED 1
// will blink once per second since that pin is also
// used to send text to the Serial Monitor
// (Click on the Arduino IDE Serial Monitor button)
// Also, in this scenario you get a do-over with your
// missile launch, rather than starting WWIII.


// Allocate switches
int switchOnePin = 8;
int switchTwoPin = 9;

// Set launch pin
int launchPin = 7;

// preset switch state
boolean switchOneState = false;
boolean switchTwoState = false;

void setup()
{

  // Set serial link to 9600 baud
  Serial.begin(9600);

  // Identify yourself
  Serial.println("Launch Control Ready");
  // Init pins for input
  pinMode(switchOnePin, INPUT); // DIP 0
  digitalWrite(switchOnePin,HIGH); // Turn on pullup
  pinMode(switchTwoPin, INPUT); // DIP 1
  digitalWrite(switchTwoPin,HIGH); // Turn on pullup

  // Init pins for output
  pinMode(launchPin, OUTPUT);

  // Preset to turn LEDs off
  digitalWrite(launchPin,HIGH);

}
```

Chapter 7: Playing with software – Part 2

```
void loop()
{
  switchOneState = digitalRead(switchOnePin);
  switchTwoState = digitalRead(switchTwoPin);

  // Launch if both switches are true
  if(switchOneState && switchTwoState)
  {
    Serial.println("Launch the missiles!");
    digitalWrite(launchPin,LOW); // light up the LED
  }
  else
  {
    Serial.println("Don't launch missiles!");
    digitalWrite(launchPin,HIGH); // turn off the LED
  }
  delay(1000); //Wait a second and check again
}
```

Shift Operators

The shift operators can be used to radically speed up multiplication and division if you divide by numbers that are a power of 2 (2,4,8,16,32…). You might want to do this for tasks like averaging ADC (Analog to Digital Converter) readings. When you study ADC you'll see that sometimes you can get more accurate results if you take a bunch of readings and average them. My first inclination was to take 10 readings and then divide by 10. I chose 10 because I've got that many fingers, but if I had chosen to divide by 8 or 16 then my division would be much faster because I can use the right shift operator: >>. If I divide by 10 the compiler has to call a large and complex division function and floating-point data types (usually adding a couple of thousand bytes to the program). However, if I divide by 8 it only needs to shift bits 3 positions to the right. Three quick right shift operations versus lots of time and program space – the tradeoff is precision – those three bits shifted to the right are lost.

Let's say my readings were:

54, 62, 59, 57, 60, 59, 56, 63 = 470

The sum is 470 and 470 / 8 = 58.75

Chapter 7: Playing with software – Part 2

Remember that the largest decimal number that fits in an 8-bit byte is 256, so we must store 470 in a 16-bit integer: 0000000111010110.

```
myTotal = 470

   bit # FEDCBA9876543210 (in Hex)
myTotal 0000000111010110
```

If we shift this right three times it becomes:

```
myAverage = (myTotal >> 3);

bit # FEDCBA9876543210
 >> 3 0000000000111010110
```

The low three bits 110 fall out of the AVR and have to be swept up later. [You didn't believe that did you? They actually just disappear.] Anyway the value now (ignoring the leading zeros) is 111010 which is decimal 58.

Note that 58 is not 58.75 but you did save both time and program space over dividing by 8, so you have to decide which is better to use. The binary averaging gets you closer than all but two of the 8 readings (59). And it does it lightning fast. So if you are time constrained the tradeoff should be obvious. Besides, if your ADC reading is +- 2% accurate, then the .75 is meaningless anyway.

Masks and Macros: Using Named Bits

You probably didn't know this, but Cylon Eyes don't just sweep a single LED back and forth at one speed. No, when Cylons get excited the sweep speeds up and if they get really excited they sweep more LEDs. When they are feeling contrary they can invert the pattern. When they get walloped up side the head, they do this weird walleye sweep. When they get really angry, the LEDs vibrate. When they get confused they do an ant sweep. When they see an actual chrome toaster – like with bread in it - they blink all LEDs on and off. And when they get infected with Microsoft Windows© they generate random dots until reset. Finally, the sweep has 2 speeds: fast and faster.

We have seven patterns and we can encode them with three switches as follows:

Chapter 7: Playing with software – Part 2

```
Binary  Decimal   Pattern Select
000     0         cylonEyes
001     1         cylonEyes2
010     2         cylonEyes3
011     3         wallEyes
100     4         antEyes
101     5         vibroEyes
110     6         blinkinEyes
111     7         randomEyes
```

If you want to get your Cylon Optometry Doctorate, you have to figure out how to encode those seven patterns, the 2 speeds, and the polarity using just 5 switches. We will use bitwise operators, masks, and macros to do exactly this for our CylonOptometry project. The hardware for this project is the same as that for the Figure 33: DIP switch and LEDs drawing (see also Figures 34 and 35).

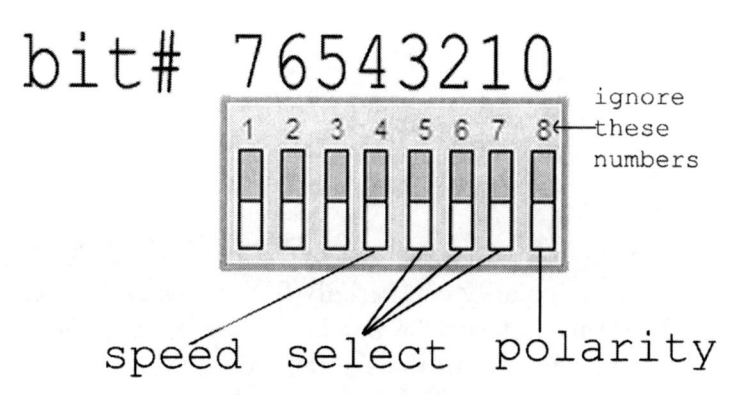

Figure 61: DIP Switch Use

Figure 61 shows that we will encode the speed and polarity with single pins and that we will use the pins 1, 2, 3 for the pattern. Note that the DIP is numbered left to right 1 to 8, but that we use bits in bytes numbered from right to left as bits 7,6,5,4,3,2,1,0.

We will define bit masks for each bit or bit group:

```
// DIP switch masks
#define POLARITYMASK    0x01    // 00000001
#define SELECTMASK      0x0E    // 00001110
#define SPEEDMASK       0x10    // 00010000
```

Chapter 7: Playing with software – Part 2

Suppose, for instance, that we set the speed to 1, the pattern select to 5, and the polarity to 1, we would see 00011011 (0x1B).

```
          PORTB = 00011011  (0x1B)

                 bit #   76543210
               PORTB  =  00011011  =  0x1B
       POLARITYMASK  =  00000001  =  0x01
                        --------
                 AND  =  00000001  =  0x01

                 bit #   76543210
               PORTB  =  00011011  =  0x1B
         SELECTMASK  =  00001110  =  0x0E
                        --------
                 AND  =  00001010  =  0x0A

                 bit #   76543210
               PORTB  =  00011011  =  0x1B
          SPEEDMASK  =  00010000  =  0x10
                        --------
                 AND  =  00010000  =  0x10
```

We have isolated the bit fields for each mask, but we still have one more step. You will note that the polarity can be only 0 or 1, which is fine, but the pattern select does not directly indicate the number of the pattern since it begins at the second bit, not the first bit. Our count for the pattern selection is not 0,1,2,3,4,5,6,7 it is 0,2,4,6,8,10,12,14. We could use either numeric sequence to define our pattern states, but it would be simpler if we could just shift the whole byte one position to the right, dropping the first bit (polarity) into the void leaving us only the pattern data – and we can do that with the right shift operator >>. Our mask gave us: 00001010 = 0x0A. But (00001010 >> 1) [shift right one bit] is equal to 00000101 so hex 0x0A >> 1 equals 0x05. The same idea holds true for the speed value that we right shift 4: (00010000 >> 4) is equal 00000001. [We could also just mask for the speed and do a true/false test and if true then the speed value is 1, if false then it is 0.]

In the CylonOptometry program we read the switch state, mask off the polarity, speed, and pattern, shift them and then use a switch statement (a decision making

Chapter 7: Playing with software – Part 2

technique that has nothing to do with the DIP switch) to select the function for the specified pattern – you can see this in the loop() function. Each of these seven functions for the LED pattern runs through an array containing the pattern to show on the LED and calls the dillydally() function that delays a number of milliseconds depending on the speed setting. It then checks to see if PORTB has changed. If not, it returns 0 to the pattern function that will show the next pattern in the array. If PORTB has changed then it returns 1 and the function will return to the loop() function and run the switch statement again to see if a new pattern has been selected.

CylonOptometery Source Code:

```
// CylonOptometry
// Joe Pardue September 16, 2008
// Redux April 12, 2009

// Function declarations
void initialize(void);
uint8_t dillyDally(void);
void cylonEyes(void);
void cylonEyes2(void);
void cylonEyes3(void);
void wallEyes(void);
void antEyes(void);
void vibroEyes(void);
void blinkinEyes(void);
void randomEyes(void);
void errorEyes(void);

// LED pattern function selection numbers
#define CYLONEYES       0x00 // 000
#define CYLONEYES2      0x01 // 001
#define CYLONEYES3      0x02 // 010
#define WALLEYES        0x03 // 011
#define ANTEYES         0x04 // 100
#define VIBROEYES       0x05 // 101
#define BLINKINEYES     0x06 // 110
#define RANDOMEYES      0x07 // 111

// DIP switch masks
#define POLARITYMASK    0x01 // 00000001
#define SELECTMASK      0x0E // 00001110
#define SPEEDMASK       0x10 // 00010000
```

Chapter 7: Playing with software – Part 2

```c
// macro to clear all bits except those
// set to 1 in the mask
#define maskBit(p,m) (p & m)

// bit 5 of DipSwitch
uint8_t volatile SweepSelect = 255;
// bit 0 of DipSwitch
uint8_t volatile Polarity = 1;

// get it started
uint8_t start = 1;

void setup()
{
  // set port B for input
  DDRB = 0x00;
  // enable pull up resistors on input port
  PORTB = 0xFF;
  // set port D for output
  DDRD = 0xFF;

  // sets up selections
  dillyDally();
}

void loop()
{
  switch (SweepSelect)
  {
      case CYLONEYES:
            cylonEyes();
            break;
      case CYLONEYES2:
            cylonEyes2();
            break;
      case CYLONEYES3:
            cylonEyes3();
            break;
      case WALLEYES:
            wallEyes();
            break;
      case ANTEYES:
            antEyes();
            break;
      case VIBROEYES:
            vibroEyes();
```

Chapter 7: Playing with software – Part 2

```
                break;
        case BLINKINEYES:
                blinkinEyes();
                break;
        case RANDOMEYES:
                randomEyes();
                break;
        default:
                errorEyes();
                break;
    }
}

// waits for 50 x milliseconds specified in SweepSpeed
// returns 0 if DipSwitch has changed
// otherwise returns 0
uint8_t dillyDally()
{
        // state of all the bits
        uint8_t static DipSwitch;
        // bits 4,5,6, and 7 of DipSwitch
        uint8_t static SweepSpeed;

        // declare and initialize the scroll delay_count
        uint32_t delay_count = 0;

        // load value
        DipSwitch = PINB;

        // load the delay count from the switch
        delay_count = (uint32_t)SweepSpeed;
        delay_count *= 50;

        // wait around for a while
          delay(delay_count);

        // check the switch to see if it has changed
        if( (DipSwitch != PINB) || start == 1 )
        {
                 start = 0;
                // polarity
                if(maskBit(PINB,POLARITYMASK)) Polarity = 1;
                else Polarity = 0;

                // mask off sweep selection
                SweepSelect = maskBit(PINB,SELECTMASK);
```

Chapter 7: Playing with software – Part 2

```
                SweepSelect = (SweepSelect >> 1);

                // mask off speed
                SweepSpeed = maskBit(PINB,SPEEDMASK);
                SweepSpeed = (SweepSpeed >> 4) + 1;

                // load new value
                DipSwitch = PINB;

                // it has changed so inform caller to bail out
                return 1;
        }
        return 0;
}
/*
00000001 == 0x01
00000010 == 0x02
00000100 == 0x04
00001000 == 0x08
00010000 == 0x10
00100000 == 0x20
01000000 == 0x40
10000000 == 0x80
01000000 == 0x40
00100000 == 0x20
00010000 == 0x10
00001000 == 0x08
00000100 == 0x04
00000010 == 0x02
*/
void cylonEyes()
{
        uint8_t i = 0;
        uint8_t ce[] = {0x01,0x02,0x04,0x08,0x10,0x20,0x40,0x80};

        while(1)
        {
                // run up the array
                for(i = 0; i <= 7; i++)
                {
                        if(dillyDally()) return; // delay or bail out

                        // show non-inverted
                        if(Polarity) PORTD = ce[i];
                        else PORTD = ~ce[i]; // show inverted
```

Chapter 7: Playing with software – Part 2

```c
            }
            // run down the array
            for(i = 6; i >= 1; i--)
            {
                    if(dillyDally()) return; // delay or bail out

                    // show non-inverted
                    if(Polarity) PORTD = ce[i];
                    else PORTD = ~ce[i]; // show inverted
            }
        }
}
/*
cylonEyes2
00000011 == 0x03
00000110 == 0x06
00001100 == 0x0C
00011000 == 0x18
00110000 == 0x30
01100000 == 0x60
11000000 == 0xC0
01100000 == 0x60
00110000 == 0x30
00011000 == 0x18
00001100 == 0x0C
00000110 == 0x06
00000011 == 0x03
*/
void cylonEyes2()
{
        uint8_t i = 0;
        uint8_t ce2[] = {0x03,0x06,0x0C,0x18,0x30,0x60,0xC0};

        while(1)
        {
                // run up the array
                for(i = 0; i <= 6; i++)
                {
                        if(dillyDally()) return; // delay or bail out
                        // show non-inverted
                        if(Polarity) PORTD = ce2[i];
                        else PORTD = ~ce2[i]; // show inverted
                }
                // run down the array
                for(i = 5; i >= 1; i--)
                {
```

123

Chapter 7: Playing with software – Part 2

```
                if(dillyDally()) return; // delay or bail out
                // show non-inverted
                if(Polarity) PORTD = ce2[i];
                else PORTD = ~ce2[i]; // show inverted
            }
        }
}
/*
cylonEyes3
00000111 == 0x07
00001110 == 0x0E
00011100 == 0x1C
00111000 == 0x38
01110000 == 0x70
11100000 == 0xE0
01110000 == 0x70
00111000 == 0x31
00011100 == 0x1C
00001110 == 0x0E
00000111 == 0x07
*/
void cylonEyes3()
{
    uint8_t i = 0;
    uint8_t ce3[] = { 0x07, 0x0E, 0x1C, 0x38, 0x70, 0xE0 };

    while(1)
    {
        // run up the array
        for(i = 0; i <= 5; i++)
        {
            if(dillyDally()) return; // delay or bail out

            // show non-inverted
            if(Polarity) PORTD = ce3[i];
            else PORTD = ~ce3[i]; // show inverted
        }
        // run down the array
        for(i = 4; i >= 1; i--)
        {
            if(dillyDally()) return; // delay or bail out

            // show non-inverted
            if(Polarity) PORTD = ce3[i];
            else PORTD = ~ce3[i]; // show inverted
```

Chapter 7: Playing with software – Part 2

```
                }
        }
}

/*
wallEyes
00000000 == 0x00
00011000 == 0x18
00100100 == 0x24
01000010 == 0x42
10000001 == 0x81
01000010 == 0x42
00100100 == 0x24
00011000 == 0x18
*/
void wallEyes()
{
        uint8_t i = 0;
        uint8_t we[] = { 0x00, 0x18, 0x24, 0x42, 0x81 };

        while(1)
        {
                // run up the array
                for(i = 0; i <= 4; i++)
                {
                        if(dillyDally()) return; // delay or bail out

                        // show non-inverted
                        if(Polarity) PORTD = we[i];
                        else PORTD = ~we[i]; // show inverted
                }
                // run down the array
                for(i = 3; i >= 1; i--)
                {
                        if(dillyDally()) return; // delay or bail out

                        // show non-inverted
                        if(Polarity) PORTD = we[i];
                        else PORTD = ~we[i]; // show inverted
                }
        }
}

/*
antEyes
01001001 == 0x49
```

Chapter 7: Playing with software – Part 2

```
10010010 == 0x92
00100100 == 0x24
*/
void antEyes()
{
      uint8_t i = 0;
      uint8_t ae[] = { 0x049, 0x92, 0x24 };

      while(1)
      {
            // loop thru the array
            for(i = 0; i <= 2; i++)
            {
                  if(dillyDally()) return; // delay or bail out

                  // show non-inverted
                  if(Polarity) PORTD = ae[i];
                  else PORTD = ~ae[i]; // show inverted
            }
      }
}

// Notice that Polarity doesn't matter for the next three patterns

void vibroEyes()
{
      while(1)
      {
            if(dillyDally()) return; // delay or bail out
            PORTD = 0xAA; // 10101010
            if(dillyDally()) return; // delay or bail out
            PORTD = 0x55; // 01010101
      }
}

void blinkinEyes()
{
      while(1)
      {
            if(dillyDally()) return; // delay or bail out
            PORTD = 0x00; // 00000000
            if(dillyDally()) return; // delay or bail out
            PORTD = 0xFF; // 11111111
      }
}
```

Chapter 7: Playing with software – Part 2

```
void randomEyes()
{
    while(1)
    {
        if(dillyDally()) return; // delay or bail out
        PORTD = rand();
    }
}

// no movement, just stares at you till you fix the error
// you won't see this, but I saw it a lot during development
void errorEyes()
{
    while(1)
    {
        // 00011000 show error state
        PORTD = 0x18;
        if(dillyDally()) return; // delay or bail out
    }
}
```

Looking carefully at this code you may notice that there seems to be a lot of repetition of similar parts, if you typed this all in to test it you may have found yourself using cut and paste a lot - and you might think: 'you know, with a little thought about creating some additional functions to handle the repetitious parts, this could actually be shorter and more efficient." And you'd be exactly right. But at this point our goal is to learn about bitwise operators, not clever software practices so for many folks it may be better to have some redundant code and less confusion.

Chapter 8: Communicating with a PC

We have already learned a little about types of data. You may remember that the **byte** (8-bits - can hold values from 0 to 255) is the basic data type that the AVR uses to store and retrieve data. It also uses the **integer** (16-bits) to handle the addresses of data, giving it unique 65535 locations to store those bytes. [Under the hood it can be more complex] Our goal in this section is to learn about sending data to and receiving data from a terminal program on the PC, so we need to learn some more about that process.

ASCII and the char data type

In Chapter 7: Playing with software – Part 2, we learned about bits and bytes and why there are exactly 10 types of people in the world. Now we will look at another data type, the char, which like a byte is also 8-bits, but was defined for a quite different set of uses. The keyword: **char** is short for **char**acter and is used primarily to manipulate character data. [Char is pronounced exactly like those first four letters in the word 'character' though the majority of engineers seem to think it is pronounced like the first four letters in 'charcoal', some even think it is pronounced like 'car' – so don't be surprised how it is pronounced in the highly unlikely event that this comes up in conversation.] The char data type usually represents information coded in ASCII (and of course there are exceptions). You can see a chart of the ASCII codes in Appendix 1: ASCII Table. The first thing you may note is that most of the codes are for printable characters like those in the text you are reading, but many with lower values, 0 to 30, have odd looking names, like 'bel' and 'nl'. These codes are legacy from the time when ASCII was first standardized to provide codes for TeleType printers. If you sent 'bel' the TeleType would ding a bell, if 'nl' it would advance the paper to a new line. The coding convention remained after the printer was replaced by a video terminal.

You may have also noticed that even though a char data type is 8-bits which could encode 256 items, 0 to 255, ASCII only encodes half that amount, 128 items (0 to 127). This is because ASCII is a 7-bit code. Over time there have been various extensions to ASCII to fill up the other slots, and now text is often encoded with

16 bits to accommodate languages other than English, but for our communication purposes, 128 characters is more than enough.

One of the examples that comes with Arduino: File\Sketchbook\Examples\Communications\ASCIITable will send the printable ASCII characters, 33 to 126 to a PC terminal program as the character, the decimal, hexadecimal, octal, and binary equivalents.

So what's the big deal about ASCII anyhow?

Think a moment about sending data between a PC and an Arduino. Suppose you wanted to tell the Arduino to read data on analogInput pin 0 once per second and send that data to a PC program that will draw a graph of the values. The Arduino uses the AVR USART peripheral (serial port) to receive data as 8-bit bytes so we can just send a stream of byte values from 0 to 255 representing our reading – right? Well, how does the PC know what it is receiving? Well, we could send it the characters for the word 'start' and write the PC program such that after it sees the character sequence for 'start' every subsequent byte it gets is raw data – right? Well, how do we tell it when we are through sending data? We can write the PC program so that if it sees the word 'stop' it stops graphing the data – right? Well, how will it tell the difference between the data bytes: 115, 116, 111, and 112 and the word 'stop' which is coded in ASCII with those values? Okay, we can agree to send 'start' and then send exactly 1000 data bytes and then after that anything else is another instruction... Well... you are beginning to see a pattern here: how can you tell the difference between coded text messages and raw data? And one answer is that you don't send raw data values, you send everything in ASCII coded characters. If you have a data point with a value of 115, you don't send that raw byte, but three bytes: 49, 49, and 53, which are the ASCII encoded characters for '1','1', and '5'. This may seem like a lot of extra hassle. Not only do you have to send three bytes instead of one, you have to convert the raw byte into the characters and then on the other end convert those characters back into a raw byte. Well, it may seem like a hassle, but it can solve a lot of problems so this is a very common way to send data using a serial port and this is what we will use.

Sending an receiving numbers

Sending numbers from the Arduino is really simple using the built-in serial functions. Receiving numbers is not so simple. Let's start by receiving some byte

Chapter 8: Communicating with a PC

numbers on the Arduino and then interpret those as commands. Later we will expand on this and start receiving char values as text commands.

Let's write a program, Number_Commander that receives an ASCII encoded character for a single digit and based on that digit, calls a function specific to that number. The user opens a terminal program, such as the Serial Monitor in the Arduino IDE (Figure 63: Number Commander in the Serial Monitor) Then sends a digit, in our example case from 0 or 1 or 2 or 3 or 4, to call one of the cmd#() functions (where # is 0 to 4). Sending any other number will cause the program to send an error message. In this example code each command function simply sends back a string noting that the command has been received. Keep in mind that we are not sending the numeric value, but the character representing the value.

[It will probably help if you refer to the source code cmdParse() function in the Number_Commander below while reading this paragraph.] Our command parser (divides something into its components) uses a C programming language **switch** statement that takes the character for the command number as a parameter and then looks through a list of **case** statements to find the case corresponding to the command number. For each case there is a list of things to do if that particular case is the correct one. We use this to call a function associated with the particular command and when the function returns, we call 'break' so that the code exits the switch statement (no need to look further down the list since we have already run the case it was looking for). If the switch statement gets to the bottom of the case list and still hasn't found the correct case, then a 'default' case is run, which for our example tells the user that an bad command was sent and shows that invalid number.

In C, there are two main methods for deciding among a list of possibilities, the switch/case or the if/else constructions. Some folks want to know why C needs both or how to decide when to use which one. If you want to start an argument, bring this up on any C related forum. My rule of thumb is that if you have fewer than four choices, the if/else will probably be the best. If you have four or more choices then the switch will probably be best, but the only real way to decide this is to do both and then look at the compiler output and see the assembly code generated. Different C compilers might handle the same code differently. For our purposes either way would suffice since we aren't code-size or speed constrained, but I tend to use the switch statement because it just looks better to me.

Chapter 8: Communicating with a PC

Number_Commander Source Code:

```
// Number_Commander Joe Pardue May 11, 2009

int cmd = 0;

void setup()
{
  Serial.begin(9600);
}

void loop()
{
  // check if data has been sent from the computer
  if (Serial.available()) {
    cmdParse();
  }
}

void cmdParse()
{
  cmd = Serial.read();

  switch(cmd)
  {
    case '0':
      cmd0();
      break;
    case '1':
      cmd1();
      break;
    case '2':
      cmd2();
      break;
    case '3':
      cmd3();
      break;
    case '4':
      cmd4();
      break;
    default:
      Serial.print("Invalid number command: ");
      Serial.println(cmd, BYTE);
      break;
  }
}
```

Chapter 8: Communicating with a PC

```
void cmd0()
{
  Serial.println("cmd0");
}

void cmd1()
{
  Serial.println("cmd1");
}

void cmd2()
{
  Serial.println("cmd2");
}

void cmd3()
{
  Serial.println("cmd3");
}

void cmd4()
{
  Serial.println("cmd4");
}
```

Compile and upload the sketch, then click on the Arduino IDE Serial Monitor icon as shown in Figure 62. You should observe the results shown in Figure 63.

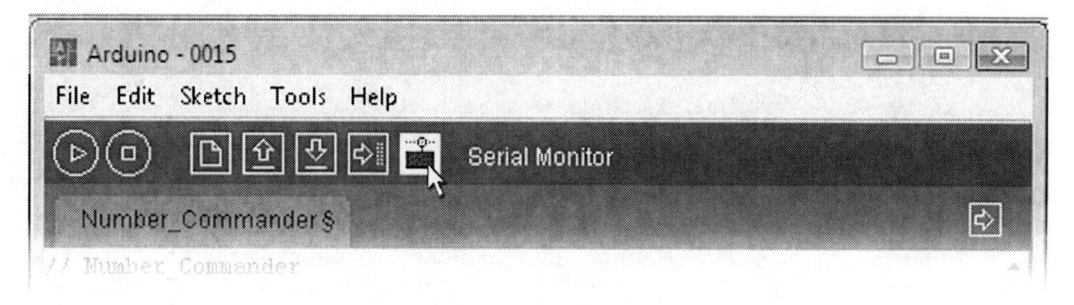

Figure 62: Select the Arduino IDE Serial Monitor

Chapter 8: Communicating with a PC

Figure 63: Number Commander in the Serial Monitor

In the Chapter 5: Some Simple Projects - Output sound - piezo music, we set up some hardware and played Happy Birthday on a piezo element glued to the bottom of a Dixie cup. And, oh boy, was the sound crappy. Now we are going to apply some of our communications learning to creating an Arduino Jukebox that will allow us to select among a bunch of crappy tunes and noises, remembering all the while that our goal here is not to go into competition with Wurlitzer, but to learn some software/hardware principles. The hardware is the same as before.

Arduino Jukebox Source Code:

```
/* Arduino Jukebox
 * Joe Pardue October 9, 2009
 * based on Arduino example code Melody
 * http://www.arduino.cc/en/Tutorial/Melody
 * (cleft) 2005 D. Cuartielles for K3
 */
// Note to more experienced programmers - yes I know this
// isn't the 'best' way to do this, but this is instructional
// code for novices.
//
// Each of the tunes repeats sections of code
// that could be combined for more economical use of resources
// but such combination might obscure things for novices.
// Eventually, we would use pointers to the tune arrays
```

Chapter 8: Communicating with a PC

```
// and play them within a single function. We would also store
// the tune data in AVR program memory.

// The following is copied from D. Cuartielles Melody example:
// BEGIN COPY
/*
 * This example uses a piezo speaker to play melodies.  It sends
 * a square wave of the appropriate frequency to the piezo,
 * generating the corresponding tone.
 *
 * The calculation of the tones is made following the
 * mathematical operation:
 *
 *       timeHigh = period / 2 = 1 / (2 * toneFrequency)
 *
 * where the different tones are described as in the table:
 *
 * note        frequency     period        timeHigh
 * c              261 Hz     3830          1915
 * d              294 Hz     3400          1700
 * e              329 Hz     3038          1519
 * f              349 Hz     2864          1432
 * g              392 Hz     2550          1275
 * a              440 Hz     2272          1136
 * b              493 Hz     2028          1014
 * C              523 Hz     1912          956
 *
 */
// END COPY

// define numbers for tunes
#define Twinkle 0
#define Happy_Birthday 1
#define Euro_Siren 2
#define Twirdle 3
#define Beep_Beep 4

// create and intialize global variables
int speakerPin = 9; // pin to drive the piezo element
int tune = 0; // tune to play

void setup() {
  Serial.begin(9600);
  pinMode(speakerPin, OUTPUT);

  // greetings
```

Chapter 8: Communicating with a PC

```
  Serial.println("TAW Tunes");
  Serial.println("Enter 0 for Twinkle Twinkle Little Star");
  Serial.println("Enter 1 for Happy Birthday");
  Serial.println("Enter 2 for Euro Siren");
  Serial.println("Enter 3 for Twerdle Alarm");
  Serial.println("Enter 4 for Beep Alarm");
}

void loop() {
  // check if data has been sent from the computer
  if (Serial.available()) {
    cmdParse(); // if true, get the data and parse it
  }
}

// use a switch statement to decide which tune to play
void cmdParse()
{
  tune = Serial.read();

  switch(tune)
  {
    case '0':
      play_Twinkle();
      break;
    case '1':
      play_Happy();
      break;
    case '2':
      play_Euro();
      break;
    case '3':
      play_Twerdle();
      break;
    case '4':
      play_Beep();
      break;
    default:
      Serial.print("TAW Invalid tune: ");
      Serial.println(tune, BYTE);
      break;
  }
}

void playTone(int tone, int duration) {
  for (long i = 0; i < duration * 1000L; i += tone * 2) {
```

Chapter 8: Communicating with a PC

```
    digitalWrite(speakerPin, HIGH);
    delayMicroseconds(tone);
    digitalWrite(speakerPin, LOW);
    delayMicroseconds(tone);
  }
}

void playNote(char note, int duration) {
  char names[] = { 'c', 'd', 'e', 'f', 'g', 'a', 'b', 'C' };
  int tones[] = { 1915, 1700, 1519, 1432, 1275, 1136, 1014, 956
};

  // play the tone corresponding to the note name
  for (int i = 0; i < 8; i++) {
    if (names[i] == note) {
      playTone(tones[i], duration);
    }
  }
}

//Twinkle
int Twinkle_length = 15; // the number of notes
char Twinkle_notes[] = "ccggaagffeeddc "; // a space represents a rest
int Twinkle_beats[] = { 1, 1, 1, 1, 1, 1, 2, 1, 1, 1, 1, 1, 1, 2, 4 };
int Twinkle_tempo = 300;
void play_Twinkle()
{
  for (int i = 0; i < Twinkle_length; i++) {
    if (Twinkle_notes[i] == ' ') {
      delay(Twinkle_beats[i] * Twinkle_tempo); // rest
    } else {
      playNote(Twinkle_notes[i],Twinkle_beats[i]*Twinkle_tempo);
    }
    // pause between notes
    delay(Twinkle_tempo / 2);
  }
}

//Happy Birthday
int Happy_length = 26; // the number of notes
char Happy_notes[] = "ccdcfeccdcgfccCafedbbafgf "; // a space represents a rest
int Happy_beats[] = { 1, 1, 2, 2, 2, 4, 1, 1, 2, 2, 2, 4, 1, 1, 2, 2, 2, 2, 6, 1, 1, 2, 2, 2, 2, 4 };
```

Chapter 8: Communicating with a PC

```
int Happy_tempo = 150;
void play_Happy()
{
  for (int i = 0; i < Happy_length; i++) {
    if (Happy_notes[i] == ' ') {
      delay(Happy_beats[i] * Happy_tempo); // rest
    } else {
      playNote(Happy_notes[i], Happy_beats[i] * Happy_tempo);
    }
    // pause between notes
    delay(Happy_tempo / 2);
  }
}

//Euro Siren
int Euro_length = 2; // the number of notes
char Euro_notes[] = "Cc"; // a space represents a rest
int Euro_beats[] = { 3, 3 };
int Euro_tempo = 150;
void play_Euro()
{
  // play it 10 times
  for(int j = 0 ; j < 10; j++){
  for (int i = 0; i < Euro_length; i++) {
    if (Euro_notes[i] == ' ') {
      delay(Euro_beats[i] * Euro_tempo); // rest
    } else {
      playNote(Euro_notes[i], Euro_beats[i] * Euro_tempo);
    }
    // pause between notes
    delay(Euro_tempo / 2);
  }
  }
}

//Twerdle
int Twerdle_length = 16; // the number of notes
char Twerdle_notes[] = "cdefgabCCbagfedc"; // a space represents a rest
int Twerdle_beats[] = { 1,1,1,1,1,1,1,1,1,1,1,1,1,1,1,1 };
int Twerdle_tempo = 50;
void play_Twerdle()
{
  // play it 10 times
  for(int j = 0 ; j < 10; j++){
  for (int i = 0; i < Twerdle_length; i++) {
```

138

Chapter 8: Communicating with a PC

```
    if (Twerdle_notes[i] == ' ') {
      delay(Twerdle_beats[i] * Twerdle_tempo); // rest
    } else {
      playNote(Twerdle_notes[i],Twerdle_beats[i]*Twerdle_tempo);
    }
    // pause between notes
    delay(Twerdle_tempo / 2);
  }
  }
}

//Beep Beep
int Beep_length = 4; // the number of notes
char Beep_notes[] = "CC "; // a space represents a rest
int Beep_beats[] = { 4,4,1  };
int Beep_tempo = 300;
void play_Beep()
{
  // play it 10 times
  for(int j = 0 ; j < 10; j++){
  for (int i = 0; i < Beep_length; i++) {
    if (Beep_notes[i] == ' ') {
      delay(Beep_beats[i] * Beep_tempo); // rest
    } else {
      playNote(Beep_notes[i], Beep_beats[i] * Beep_tempo);
    }
    // pause between notes
    delay(Beep_tempo / 2);
  }
  }
}
```

These two programs are about getting user input. You might want to pause a moment and compare this technique, which requires a PC link via a serial port, with the earlier technique for getting user input from a DIP switch as shown for the Cylon Eyes program. There are many ways to get user input problem.

Getting real with serial input

As I said earlier the Arduino IDE provides a simple Serial Monitor (PC side serial terminal program) and some serial communications functions that allow you communicate with the Arduino board. These functions do a good job in sending serial text from the Arduino to the PC, but, IMHO, not such a hot job of receiving data from the PC. The Serial.available() function tells you when there is a character available on the serial port, and the Serial.read() function will fetch that character for you. However, the Arduino does not provide (that I know of) any Arduino-like simplified way to deal with those characters once you've got them. I contrast this weakness (IMHO) to a real strength of the C programming language: the C Standard Libraries contain a wealth of functions that help in dealing with data input and output over a serial port. I admit that you can use those C libraries with the Arduino IDE, but that kind of defeats the purpose of Arduino dun'nit?

Let's look at an Arduino program that uses logic that mimics one of those C library functions: atoi() (**a**scii **to** **i**nteger) that will allow us to input a sequence of numeric characters using the Arduino Serial Monitor and then convert those characters to an integer data value from 0 to 65535. We will use this later to set a motor speed.

The Number_Commander program allowed us to pick a tune to play by entering a numeric character in the Arduino IDE Serial Monitor. While that was cool, it limited us to 10 choices based on input of the characters from 0 to 9. Please bear with me while I repeat some concepts mentioned earlier (repetition reinforces learning, or so they say). We weren't actually looking at the numbers 0 to 9, but the ASCII character code that represent those numbers. This is an important distinction because the numeric characters are coded with values that are not the same as what we would normally think of as the value of that character. The character '0' is not coded with a value of 0, but with an ASCII code value of 48. Each subsequent numeric character: 1 to 9 is coded as 49 to 57. The ASCII values of 0 to 9 are codes for the communication device, (for instance, as I said before, the ASCII code with a value of 7 was used to ring a bell, other low numbered codes were used to do things like advance printer paper or return the print head to the left). The Arduino IDE Serial Monitor allows you to send characters from your PC keyboard, but if you want to send a real number, say 127, then we have to have some way of receiving the characters '1', '2' and '7' and some end-of-number character such as '!' so that the software can know that number sequence has ended. We write a program for the Arduino that stores numeric characters until it sees a '!', then converts those characters to the number they represent. For instance we could send six characters: 42356! and then convert those characters to an integer with a numeric value of 42,356.

Chapter 8: Communicating with a PC

Rather than spend a lot of space with further explanations, we'll just look at the program ASCII_To_Integer. Please be aware that this program can be easily spoofed with bad input, but for learning purposes it will suffice.

If you want to experience a real 'computer programming' moment, first think about how you would collect a sequence of ASCII coded numeric characters and convert them to a number - then walk through the following code with a pencil and piece of paper - especially the ATOI algorithm that extracts the number from the character string. Compare that with how you would have done it to how the C guys did it. My initial guess worked, but wasn't nearly as good as the code in the C Standard Library. The guys that came up with stuff like this were not only clever; they wrote some amazingly efficient code.

ASCII_To_Integer Source Code:
```
// ASCII_To_Integer 8/1/09 Joe Pardue
// Duplicates (somewhat) the C Standard Library
// function atoi() algorithm using Arduino.
// Note that there is no filtering of the input
// so if you enter something other than an integer
// value from 0! to 65535!, well, good luck.

int myInput = 0;
int myNum[6];
int myCount = 0;
int i = 0;
int n = 0;

void setup()
{
   Serial.begin(9600);
   Serial.println("ASCII_To_Integer");
}

void loop()
{
  // get characters until receiving '!'
  while( myInput != '!' ) getNum();

  // convert end-of-number character '!' to 0
  myInput = 0;
  myNum[--myCount] = 0;
```

Chapter 8: Communicating with a PC

```
  // convert ASCII string to integer
  ATOI();

  // clean up and do it all again
  clearAll();
}

// Put serial characters in a character array
void getNum()
{
  if(Serial.available())
  {
     myInput = Serial.read();
     // put the character in the array
     myNum[myCount++] = myInput;
  }
}

void ATOI()
{
  // algorithm from atoi() in C standard library
  n = 0;
  for(i = 0; myNum[i] >= '0' && myNum[i] <= '9'; ++i)
     n = 10 * n + (myNum[i] - '0');

  // show the number
  Serial.print("You sent: ");
  Serial.println((unsigned int)n,DEC);
}

void clearAll()
{
  myCount = 0;
  for(i = 0; i < 6; i++)
  {
    myNum[i] = 0;
  }
  Serial.flush();
}
```

Load this on your Arduino and you'll get results as shown below in Figure 64: Arduiono IDE Serial Monitor ASCII_To_Integer.

142

Chapter 8: Communicating with a PC

Figure 64: Arduiono IDE Serial Monitor ASCII_To_Integer

Command Demonstration

Text based command interpreters can get very clever very quickly, in fact they can get so clever so quick that they can be very hard to understand. We will stick to a not-so-clever, but easier to understand way of doing things.

Command Demonstration Specification:
1. Accepts input until receiving 'terminator' character, where 'terminator' is set in a #define statement. The default value will be '!' the exclamation character.
2. Parses command text as groups of 'separator' delimited characters, where 'separator' is set in a #define statement. The default value will be ',' the comma character.
3. Parses commands as nested case statements.
4. Convert a text number following 'n' to an unsigned integer.

Since this may be a bit hard to visualize, let's look first at the results of the program as shown below in Figure 65: Command Demo.

Chapter 8: Communicating with a PC

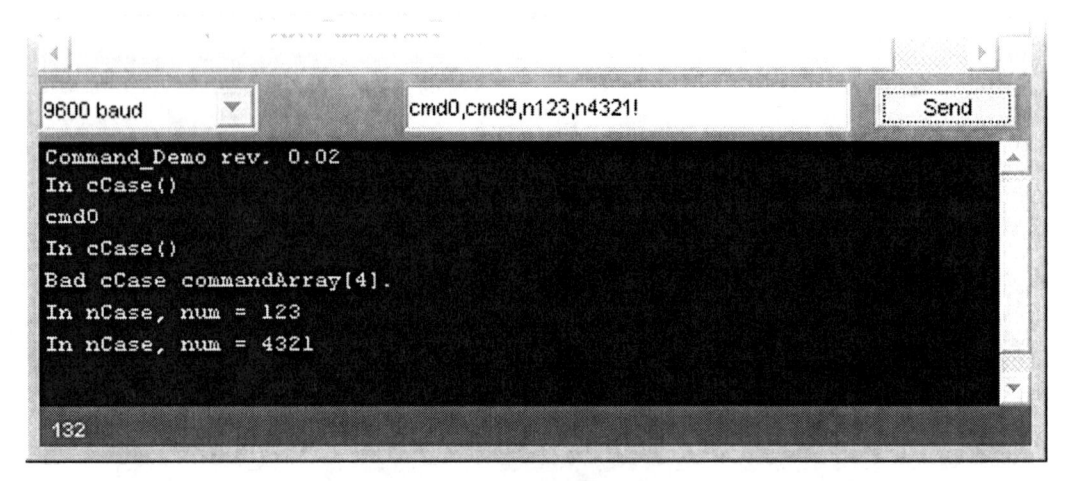

Figure 65: Command Demo

Command_Demo Source Code:

```
// Command_Demo Joe Pardue October 4, 2009
// This software demonstrates some principles in receiving
// text from a PC and interpreting that text as commands and
// numbers. It is an educational demonstration and is neither the
// best way to do things nor does it have more than cursory error
// catching.

// Define constants
#define TERMINATOR '!' // Terminator character
#define SEPARATOR ',' // Separator character
#define MAX_COMMAND_LENGTH 16
#define MAX_COMMANDS 8
#define MAX_MESSAGE MAX_COMMAND_LENGTH * MAX_COMMANDS
#define MAX_NUMBER_SIZE 6

// Create arrays
byte commandArray[MAX_COMMAND_LENGTH];
byte messageArray[MAX_MESSAGE];
int myNum[MAX_NUMBER_SIZE]; // used by ATOI

// Counts
int commandArrayCount = 0;
int messageArrayCount = 0;
```

Chapter 8: Communicating with a PC

```
void setup()
{
  // Set up the serial port
  Serial.begin(9600);

  // Say howdy
  Serial.println("Command_Demo rev. 0.02");

  initializeMessageArray();
}

void loop()
{
  if(Serial.available())
  {
    loadMessageArray();
  }
}

void loadMessageArray()
{
  if(messageArrayCount <= MAX_MESSAGE)
  {
    // put the character in the array
    messageArray[messageArrayCount] = Serial.read();
  }

  if(messageArray[messageArrayCount] == TERMINATOR)
  {
    parseArray();
  }
  else messageArrayCount++;
}

void parseArray()
{
  int i = 0;
  int j = 0;
  while( (messageArray[i] != TERMINATOR) && (i < MAX_MESSAGE) )
  {
    while( (messageArray[i] != SEPARATOR) && (messageArray[i] != TERMINATOR))
    {
      commandArray[j++] = messageArray[i++];
    }
    commandArrayCount = j;
```

Chapter 8: Communicating with a PC

```
      parseCommand();
      j = 0;
      i++;
    }
  messageArrayCount=0;
  initializeMessageArray();
}

void parseCommand()
{
  switch (commandArray[0])
  {
    case 'c':
      cCase();
      break;
    case 'n':
      nCase();
      break;
    case 'l':
      lCase();
      break;
    default:
      Serial.println("Bad commandArray[0].");
      break;
  }
}

// ASCII TO Integer
int ATOI()
{
  // algorithm from atoi() in C standard library
  int i = 0;
  int n = 0;
  for(i = 0; myNum[i] >= '0' && myNum[i] <= '9'; ++i)
    n = 10 * n + (myNum[i] - '0');

  return(n);
}

void clearMyNum()
{
  for(int i = 0; i < MAX_NUMBER_SIZE - 1; i++)
  {
    myNum[i] = 0; // set array to null
  }
}
```

Chapter 8: Communicating with a PC

```
void initializeMessageArray()
{
   // Fill with the terminator character
   for(int i = 0; i < MAX_MESSAGE; i++)
   {
     // Fill array with the terminal character
     messageArray[i] = TERMINATOR;
   }
}

// demonstrate how to extract a number
void nCase()
{
  unsigned int num = 0;

  // the number follow 'n' and is commandArrayCount long
  if(commandArrayCount <= MAX_NUMBER_SIZE)
  {
    clearMyNum();
    for(int i = 0; i < commandArrayCount-1; i++)
    {
      // skip the first element 'n' in the array
      myNum[i] = commandArray[i+1];
    }
    num = ATOI();
    Serial.print("In nCase, num = ");
    Serial.println(num,DEC);
  }
  else
  {
    Serial.println("commandArrayCount > MAX_NUMBER_SIZE");
  }
}

void cCase()
{
  Serial.println("In cCase()");

  switch (commandArray[1])
  {
    case 'm':
    if(commandArray[2] == 'd')
    {
      switch (commandArray[3])
```

Chapter 8: Communicating with a PC

```
      {
        case '0':
          Serial.println("cmd0");
          break;
        case '1':
          Serial.println("cmd1");
          break;
        case '2':
          Serial.println("cmd2");
          break;
        default:
          Serial.println("Bad cCase commandArray[4].");
          break;
      }
    }
    else Serial.println("not d");
    break;
    default:
      Serial.println("Bad cCase commandArray[1].");
      break;
  }
}

void lCase()
{
  Serial.println("lCase()");
}
```

Chapter 9: Sensors

Chapter 9: Sensors

Light Sensor one leg to +5V
other leg to 10k Ω to GND
and to Analog In pin 0

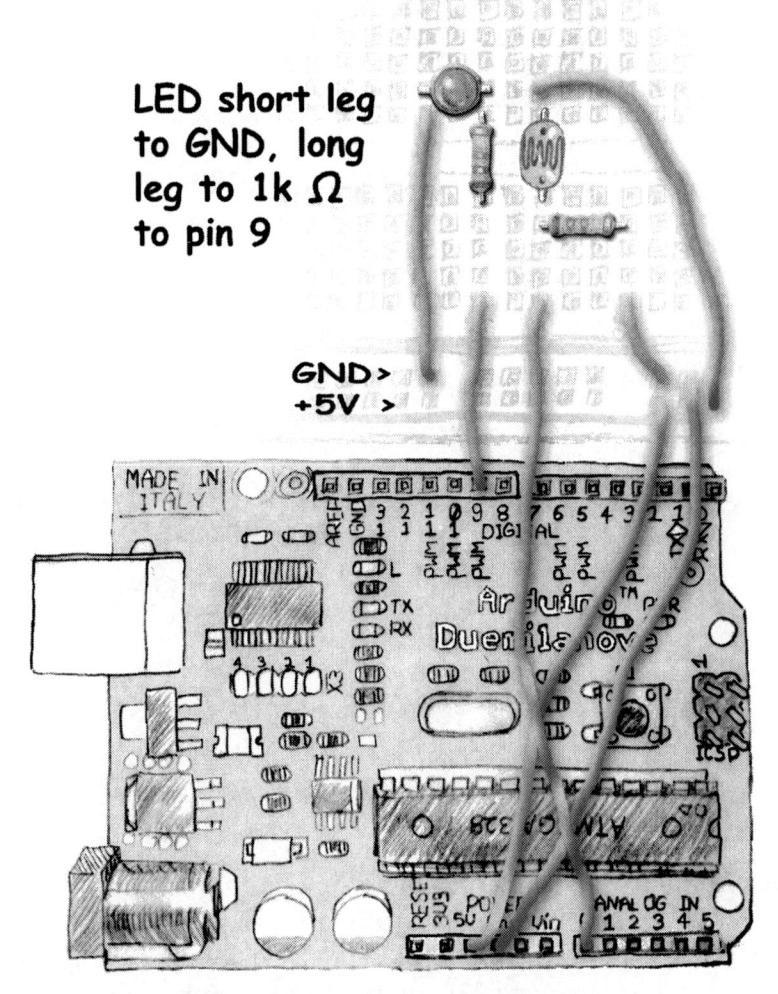

LED short leg
to GND, long
leg to 1k Ω
to pin 9

GND >
+5V >

Figure 66: CdS light sensor layout

Chapter 9: Sensors

In this chapter we will use some more of the Arduino Projects Kit parts to sense visible light, temperature, and infrared radiation. We will also learn some concepts for handling the kinds of data that we derive from sensors.

Light Sensor

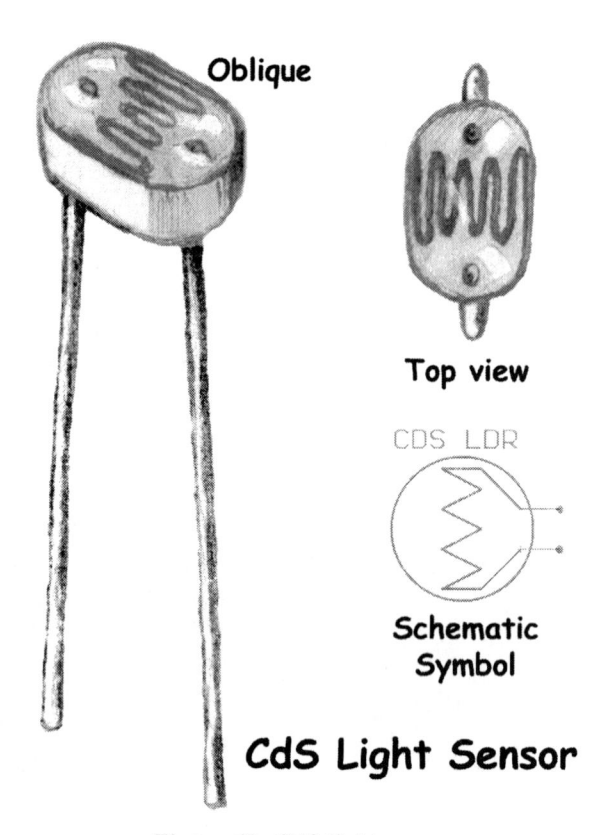

Figure 67: CdS light sensor

Cadmium Sulfide (CdS) has the interesting property that its resistance drops proportional to the amount of light falling on it. Figure 67: CdS light sensor, shows the sensor in the Arduino Projects Kit. In the dark the resistance is about 138k ohms that drops to about 23k ohms under my work lights. What you get will of course vary depending on your lights. Theoretically, you could use this to make a sensor that outputs some physical measure of light such as lumens, but the process of calibrating such a sensor to a standard brightness is far more complex

150

Chapter 9: Sensors

than we want to get into at the moment. So we will simply use our sensor to recognize changes in light in our specific environment.

The source for this code is nearly identical to the AnalogInput example in the Arduino IDE. Try not to get confused between the example and the code used for the layout in Figure 66: CdS light sensor layout. In the Arduino code the ledPin should be changed to pin 9 and the sensor pin to Analog In 0. And no, I didn't move these pins around to confuse you. I did it to confuse myself. Sorry about that. But it does show how easy it is to change the pins in either the code or the layouts just as long as you keep them consistent.

Light Sensor Components, Schematic, Layout

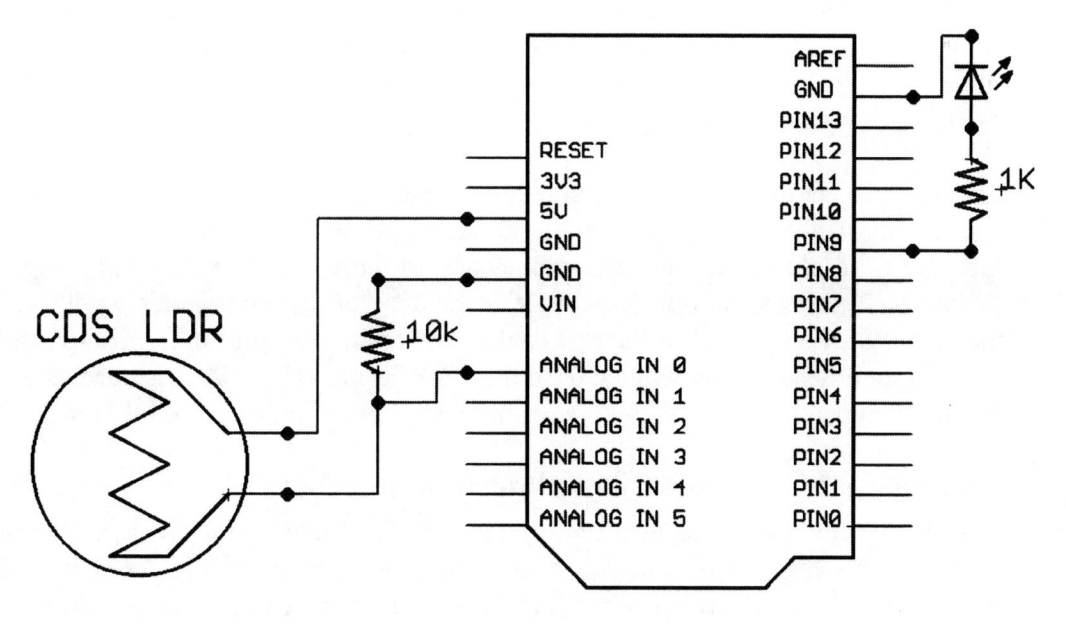

Figure 68: CdS light sensor schematic

You can follow the above schematic to build this light sensor project hardware. Then modify the Arduino example for Analog Input to conform to the actual pinout that uses pin9 and Analog in 0. With these changes, the AnalogInput example code will modify the blink rate of an LED as an indicator of relative brightness.

Chapter 9: Sensors

A Word or Two about Storing and Showing Sensor Data

We used the cop-out with the light sensor that calibrating the data to some physical reference was beyond what we want to do here (and 'we' didn't even ask for your opinion!). So we let an LED blink at a rate proportional to the sensor output from the lights in my personal workshop. We can assume that your workshop will be more or less lighted the same and that you can adjust the code if you live in a cave or on a beach. But what if we wanted to calibrate the sensor to some actual physical measure of light that contains, as real physics often does, some fractional data?

There are many situations where we don't really care if anybody ever sees any indication of a sensor output, (like the oxygen sensor in our car), since the value is being used inside an embedded system that percolates along without our help. But sometimes, such as when we are sensing room temperature, we want the sensor output to have some meaning to us.

Our next project uses the LM35 temperature sensor from the Arduino Projects Kit and outputs 10mV per degree Centigrade accurate to 0.5 °C. While that is a fun fact, we'd hardly know that the room was too darn hot based on a reading of 0.375 volts (never mind our flop-sweat as a trust-worthy sensor). We want to see that it is 99.5°F [37.5 °C for the rest of the world who can translate the output mV to °C in their head if they wanted to, but I think Fahrenheit so 99.5 is hot and 37.5 is cold] Anyway the point is that we want to see something meaningful to us.

A quick introduction to signed decimal numbers

This brings us to an issue with computers. Computers use integers (whole numbers {0,1,2,3,...}) to manipulate data but we want to see temperatures with decimal fractions: 98 won't cut it, but 98.6 is a good healthy number. Also the LM35 is accurate to 0.5°C so why waste good fractional information? In our specific case we use an ADC to measure a value expressed as a whole number between 0 and 1023, but we will be showing a temperature such as 98.6°F, with an integer part: '98' a decimal point: '.', and a decimal part: '6'. To further complicate things, our temperature scales also use negative numbers like the temperature at which CO_2 changes from a solid to a gas: −78.5 °C (−109.3 °F).

Chapter 9: Sensors

Computers store data as whole numbers and they have special hardware or software to manipulate this basic whole number data type as other data types such as negative or decimal numbers. They pretend to work with decimal numbers by storing the integer part and the decimal part as separate whole numbers. They can be told to consider a number as signed (can hold positive + and negative -) numbers by looking at the most significant bit of the data [a byte holds 0 to 255, a signed byte 'holds' −128 to +127 where the highest bit value represents the sign]. All this is by way of introduction, since going into the details can get complex quickly. We will look at a technique for keeping our data in the original ADC generated whole numbers (10-bits or 0 to 1023) while presenting that data to people in a human comprehensible format such as a signed decimal number.

Showing integer data as signed decimal fractions

Let's say we want to print the normal human body temperature in Fahrenheit, which requires a decimal fraction: 98.6 °F. The first trick we use is to keep our body temperature data stored in integers that are ten times the real value. We thus store 986 rather than 98.6. The second trick is to recover the real value for the integer and decimal fraction parts only when we want to show the data to people. And the final trick is to print these two values separately with a decimal point printed between them. We show the text 98.6 by separately printing the '98' then a decimal point '.' And then the fractional part '6'. We will use the C '/' division and '%' modulo operators to get the integer and fractional parts of the number.

The '/' division operator in C yields only the integer part of the division so in C 986/10 = 98. The '%' modulo operator yields only the remainder part of a division so in C, 986%10 = 6. We can use these operators as follows:

```
// show 986 as 98.6
whole = 986/10; // divide to get the integer
decimal = 986%10; // modulo to get the fraction

Serial.print("Temperature = ");
Serial.print(whole, DEC);
Serial.print('.');
Serial.print(decimal, DEC);
```

And the results output to the terminal would be:

```
Temperature = 98.6
```

Chapter 9: Sensors

Converting Centigrade to Fahrenheit

Let's take a moment to look at another issue related to presenting the data specific to temperature. We are storing raw integers from the ADC that map directly to Centigrade values (we'll look at the electrical and microcontroller details in a minute). And, while I have nothing against °C, I am at the moment sweating to °F. To convert this we can use the standard formula:

Fahrenheit = ((9*Centigrade)/5) + 32

This formula maps the data as shown in Figure 69: Centigrade mapped to Fahrenheit, but be sure and look at the showTemp() function shown in 'LM35_Temperature Source Code' below since I also round off the data to 0.5 to conform to the sensor accuracy.

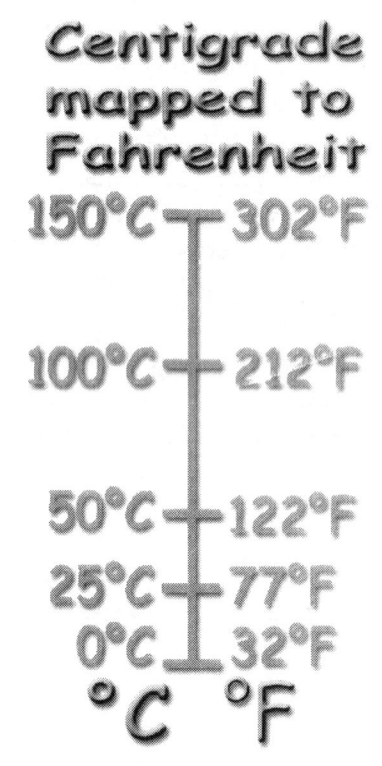

Figure 69: Centigrade mapped to Fahrenheit

Chapter 9: Sensors

We will apply this in the LM35_Temperature Source Code shown after we see how to construct the temperature sensor hardware.

The LM35 Temperature Sensor

Now that we've looked at some ideas about how to present integer data with a faked decimal point, Let's look at some more details on how to get that data.

The LM35 temperature sensor outputs voltage that is linearly proportional to the temperature in degrees Celsius (Centigrade). The LM35 datasheet says: "0.5°C accuracy guaranteeable at 25°C". It outputs 10mV per °C (5mV per 0.5°C), so if we can measure the voltage with an accuracy of 5mV, then we will match the LM35 accuracy.

Our 10-bit ADC will map an input voltage from 0 to 5 volts to integers from 0 to 1023. The resolution is 5 V / 1024 ADCunits = .0049V/ADCunit (.0049V is 4.9mV). The LM35 accurately outputs 5mV per °C and the ADC measures 4.9mV per ADCunit, so our AVR ADC is well matched with the LM35 for accuracy.

But, as we've discussed above, we don't want to mess with decimal fractions. First we note that 4.9mV is .0049 volts. We will multiply our 'per ADC unit' by 10,000 giving us the whole number 49 for each ADC unit. This is the value we will store [thus the ADCUnit constant used in the source code].

For example, if the ADC reads 204 (out of a maximum of 1023), we multiple 49*204 = 9996, which is the integer that we store. We will only extract the decimal temperature when we want to show it. So for this ADC reading the actual temperature is 100°C or 212°F as shown in Figure 70: Temperature, voltage, and ADC ranges.

Chapter 9: Sensors

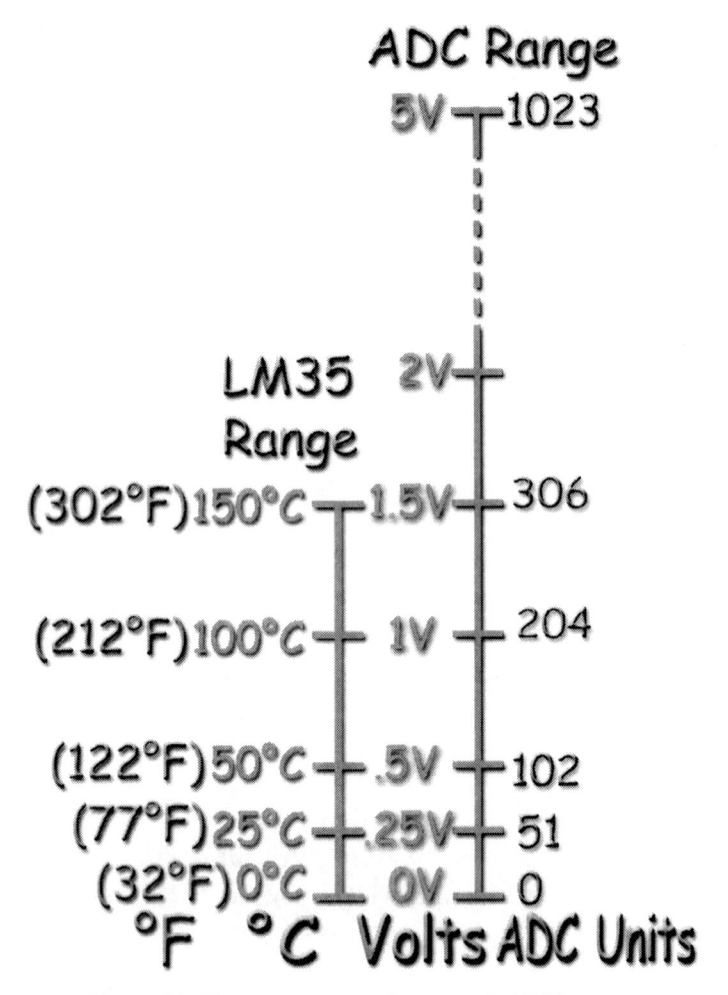

Figure 70: Temperature, voltage, and ADC ranges

Chapter 9: Sensors

Temperature Sensor Components, Schematic, Layout

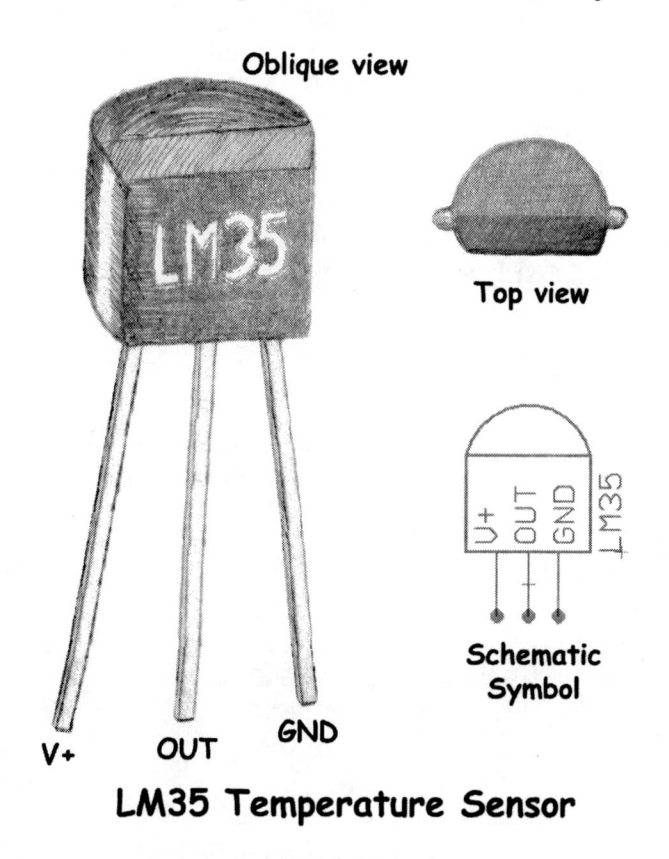

Figure 71: LM35 temperature sensor

The hardware construction is fairly simple for this project. Figure 71: LM35 temperature sensor, shows the device we will be using. Figure 72: Temperature sensor schematic, shows how we will hook it up using just three wires for +5V, GND, and Analog Pin 0 as shown in Figure 73: Temperature sensor layout drawing.

Once you get this working you can do a quick experiment to demonstrate the lag in the measured temperature and the external temperature. It takes a few seconds for the external temperature to move from the surface of the LM35 to the sensor. You can see this by observing the room temperature output. Then lightly squeeze

Chapter 9: Sensors

the LM35 between your thumb and index finger. It takes a couple of seconds to begin to respond and once you remove your fingers it takes several seconds, depending on how long you held it, for it to get back to room temperature (as shown in Figure 74: Output in Arduino Serial Monitor). In my case it took a minute or so to go from the frigid 66.0 °F (in the local bookstore where I tested it over a cup of too expensive tea) to a maximum of 85.0 °F. I don't think this low temperature indicates anything about my zombihood, since the 98.6 °F would require sticking the LM35 into a convenient orifice, something I'm not willing to do in public. (The lady at the next table is giving me strange looks, possibly because of my laptop/ALP setup and possibly because I just laughed out loud).

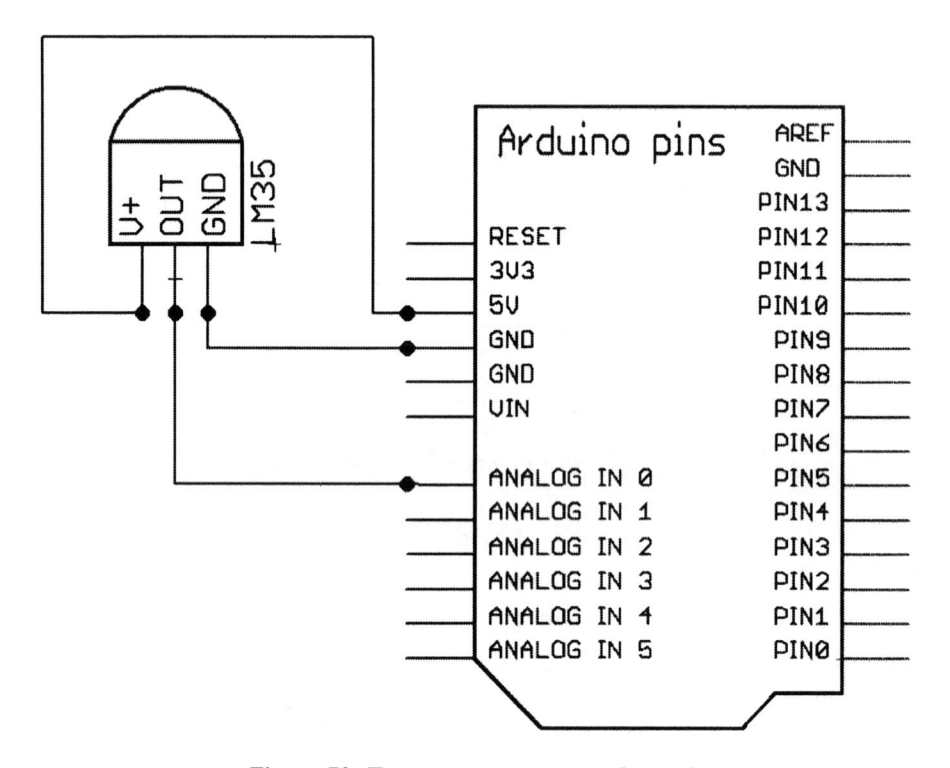

Figure 72: Temperature sensor schematic

Chapter 9: Sensors

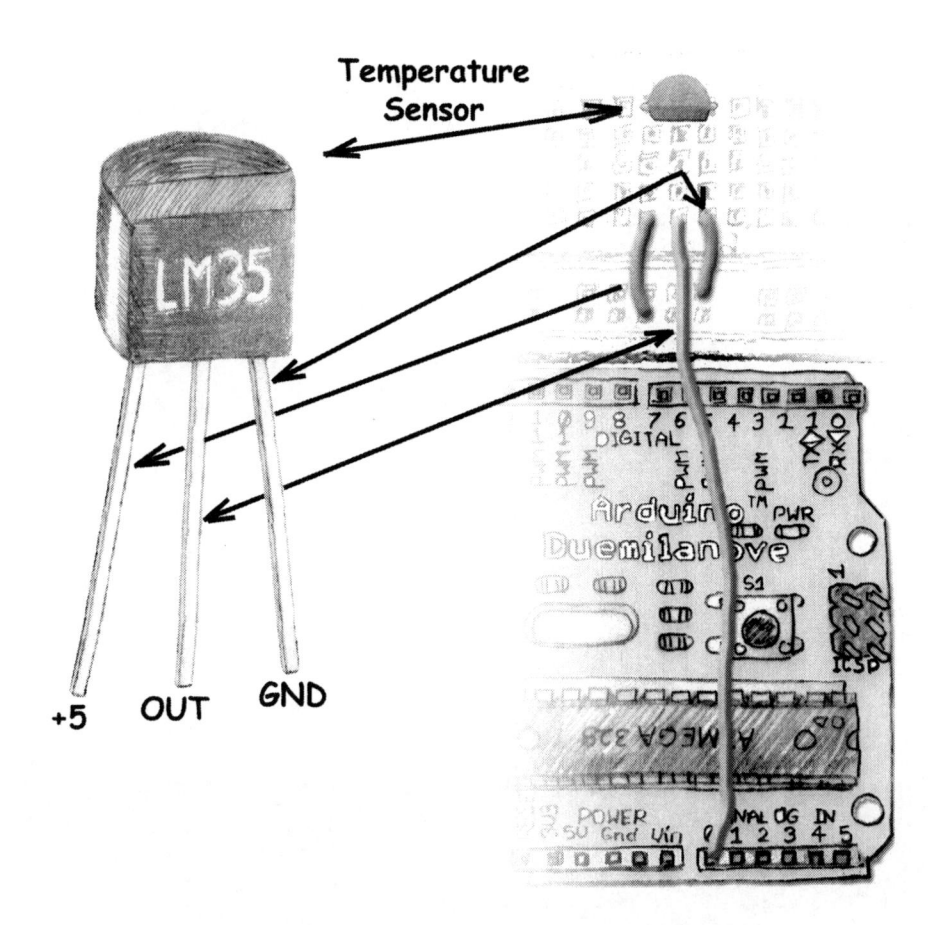

Figure 73: Temperature sensor layout drawing

LM35_Temperature Source Code:

```
// LM35_Temperature
// Joe Pardue June 4, 2009

// variable to hold the analog input value
int analogValue = 0;

// variables used to fake the decimal value
int whole = 0;
int decimal = 0;

// For the Arduino using 5V, the ADC measures
// 4.9 mV per unit use 49 to avoid floats
#define ADCUnit 49
```

159

Chapter 9: Sensors

```
void showTemp(int ADCin);

void setup()
{
  // begin the serial communication
  Serial.begin(9600);
}

void loop()
{
  // read the voltage on Analog Pin 0
  analogValue = analogRead(0);

  // show the reading with faked decimals
  showTemp(analogValue);

  // delay 1 second before the next reading:
  delay(1000);
}

void showTemp(int ADCValue)
{
  // print ADC value
  Serial.print("1.06 TAW LM35 - raw ADC: ");
  Serial.print(ADCValue, DEC);

  // make Centigrade
  whole = (ADCValue * ADCUnit)/100;
  decimal = (ADCValue * ADCUnit)%100;

  // round to '0.5'
  if(decimal > 50){ decimal = 0; whole +=1;}
  else decimal = 5;

  // print degrees Centigrade
  Serial.print(" > degree C: ");
  Serial.print(whole, DEC);
  Serial.print('.');
  Serial.print(decimal, DEC);

  // convert Centigrade to Fahrenheit
  whole = ((9*(ADCValue * ADCUnit))/5)/100;
  decimal = ((9*(ADCValue * ADCUnit))/5)%100;

  // round to '0.5'
```

160

Chapter 9: Sensors

```
    if(decimal > 50){ decimal = 0; whole +=1;}
    else decimal = 5;

    // print in degrees Fahrenheit
    Serial.print(" > degree F: ");
    whole += 32; // scale it to °F
    Serial.print(whole, DEC);
    Serial.print('.');
    Serial.println(decimal, DEC);
}
```

The output of this code in the Arduino IDE serial monitor is shown in Figure 5.

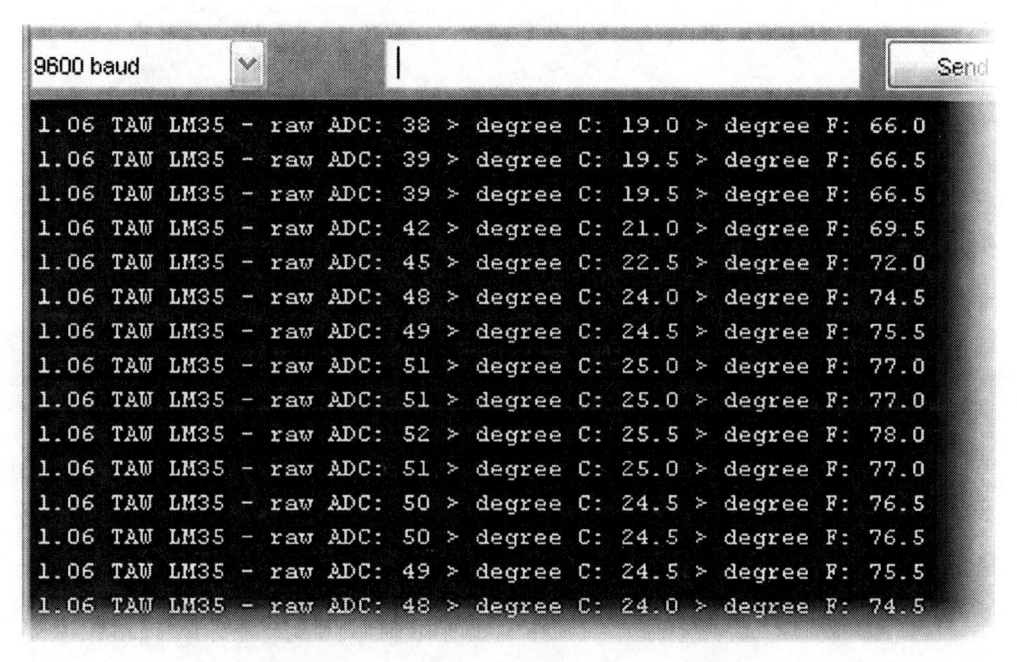

```
1.06 TAW LM35 - raw ADC: 38 > degree C: 19.0 > degree F: 66.0
1.06 TAW LM35 - raw ADC: 39 > degree C: 19.5 > degree F: 66.5
1.06 TAW LM35 - raw ADC: 39 > degree C: 19.5 > degree F: 66.5
1.06 TAW LM35 - raw ADC: 42 > degree C: 21.0 > degree F: 69.5
1.06 TAW LM35 - raw ADC: 45 > degree C: 22.5 > degree F: 72.0
1.06 TAW LM35 - raw ADC: 48 > degree C: 24.0 > degree F: 74.5
1.06 TAW LM35 - raw ADC: 49 > degree C: 24.5 > degree F: 75.5
1.06 TAW LM35 - raw ADC: 51 > degree C: 25.0 > degree F: 77.0
1.06 TAW LM35 - raw ADC: 51 > degree C: 25.0 > degree F: 77.0
1.06 TAW LM35 - raw ADC: 52 > degree C: 25.5 > degree F: 78.0
1.06 TAW LM35 - raw ADC: 51 > degree C: 25.0 > degree F: 77.0
1.06 TAW LM35 - raw ADC: 50 > degree C: 24.5 > degree F: 76.5
1.06 TAW LM35 - raw ADC: 50 > degree C: 24.5 > degree F: 76.5
1.06 TAW LM35 - raw ADC: 49 > degree C: 24.5 > degree F: 75.5
1.06 TAW LM35 - raw ADC: 48 > degree C: 24.0 > degree F: 74.5
```

Figure 74: Output in Arduino Serial Monitor

Chapter 9: Sensors

Infrared Object Detection

Figure 75: Cute L'il Bunny has her last thought.

But is it Light?

In the illustration above we see Cute L'il Bunny in the dark having her very last thought, thinking she is safe. What she doesn't know is that bad ole Mr. Snake is a pit viper and has a couple of IR Object Detectors located in pits on either side of his head between his eyes and his nose. These sensors allow him to triangulate on

162

Chapter 9: Sensors

objects that are hotter than the surrounding environment, such as bunnies that think they are well hidden in the dark, but are glowing brightly to the snake. Yum.

William Herschel accidentally discovered infrared radiation in 1800. He used a prism to split sunlight into colors and placed thermometers along the spectrum to measure the temperatures of the colors. He was shocked to note that the hottest part occurred in the unlit area just below the visible red zone, which he named infrared (infra means below). [Incidentally in 1781 he discovered Uranus, but I don't know if thermometers were involved.]

We cannot see IR with our eyes but if the source is intense we can 'see' it with our skin as the perception of warmth. You can close your eyes tightly and find a nearby heater by holding your palms out and turning around until you feel the heat – this isn't exactly high resolution 'seeing' but it is 'perceiving' IR light.

You might think it would be cool if we could see IR with our eyes. Well think about that for a moment. Since we operate at 98.6° F we, like that poor bunny, are beacons of IR light, and if we could see IR we would be blinded by the emissions in our own eyeballs. Snakes can use IR because they are cold blooded. Figure 76: Dog IR thermoscan, shows a NASA image of a dog that, in additional to looking really creepy, provides a false color view at how the dog might appear to something that can see IR. [And since the book is black and white you can't see the colors, but you can at least get the basic idea.]

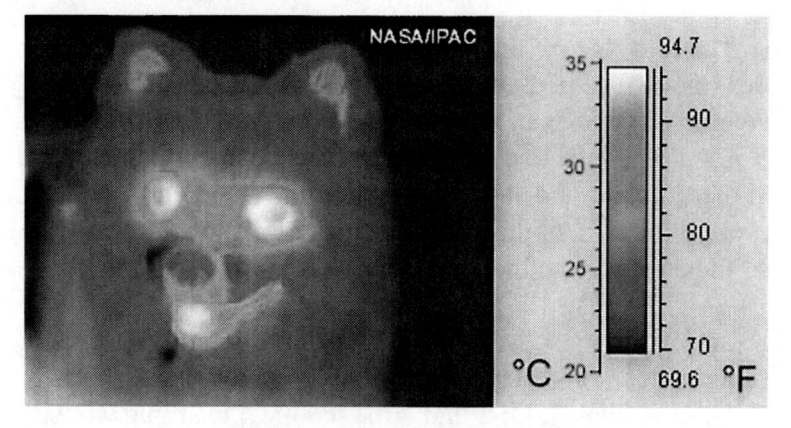

Figure 76: Dog IR thermoscan

Chapter 9: Sensors

Even though you can't see IR with your own eyes, you can detect it with a digital camera. Just hold a TV remote pointed at your camera (even a cellphone camera will show this) and press the power button. With your eyes you see nothing, but as shown in Figure 77: Digital camera 'sees' TV remote IR, the camera shows a light.

Figure 77: Digital camera 'sees' TV remote IR

Some Common Uses for IR Sensors

One of our most common electronic IR emitters is on that blessing to couch potatoes everywhere: the TV remote control. It broadcasts a coded IR signal to the IR detector on the TV console that decodes the signal and saves you the excruciating drudgery of waddling over to the TV to change the channel.

And where would the military be without IR for target acquisition? Fire a Sidewinder missile (named after a pit viper) at the exhaust pipe of your enemy's jet and say good-bye to your enemy. Or shine an IR laser pointer onto a window and drop a smart bomb that will follow the beam into that window, tap your enemy on the shoulder and say "Blooey to you Mr. Enemy". Another cool way to improve you kill skill is night vision goggles that allow your troops to be like Mr. Snake and the enemy to be like L'il Bunny.

And there are many other uses, like factory production-line sensing tomato soup cans passing on a conveyer belt, which is much more boring than blasting enemies (or eating bunnies) but that's the practical example we will construct in a few pages.

Chapter 9: Sensors

IR Reflective Object Sensor – the QRD1114

The QRD1114 is made from an IR LED emitter, an IR phototransistor detector, and an IR opaque shield that holds them together. You can remove the emitter and detector by holding the legs and pushing toward the front of the device, and if you do this, remember which way the legs were so that you can get it reassembled properly. And remember to watch for the emitter or the detector protruding beyond the black holder, as this will radically affect the way the device works. The functional concept is simple: the shielded emitter radiates IR from the open end that can only be seen by the shielded detector when the IR bounces off something, as shown in a cutaway view below.

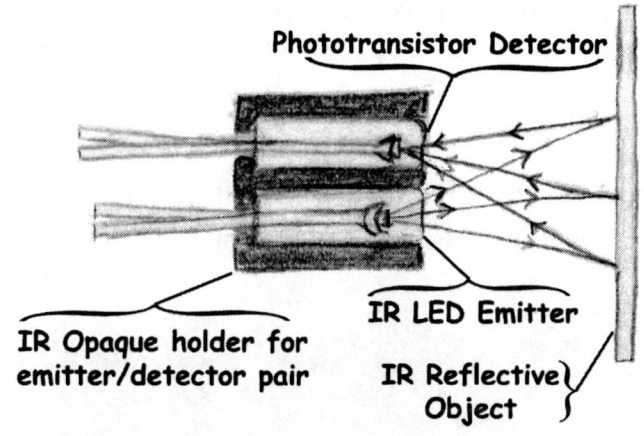

QRD1114 Cross-section

Figure 78: QRD1114 Cross-section

Making IR visible

The QRD1114 must be calibrated to a specific application to be used to its greatest effect. Calibration is necessary for two main reasons: first he sensor detects IR not just from the reflections of the associated emitter, but from ambient (environmental) IR and second, the amount of IR reflected back from the emitter varies with the IR reflectivity of the reflecting object and the distance to that object.

Chapter 9: Sensors

IR doesn't know black from white

White reflects all visible light and black absorbs it. You might think that a white area would reflect more IR back to the sensor than a black area. I did, but it turns out that the operative part of the definition of black and white is 'visible'. Some things that appear black and white to me reflect the same amount of IR and look identical to the IR sensor. I printed my first motor wheel encoder on photographic paper and my detector couldn't tell the difference in the white and black bars. Naturally, I assumed that the problem was something in the software or hardware and spent a lot of time looking for a bug, but it turns out that printing the same image on plain paper works pretty well. Apparently the shiny surface of the photo paper is transparent to the human eye but reflects the incident IR. I found that I got an even better sensor reading if I went over the plain paper printed black areas with a black Sharpie© ink pen. Again, I couldn't see much difference but the detector sure could.

Show invisible IR intensity with visible red LED

You can see a drawing of the QRD1114 along with its schematic symbol in the figure below.

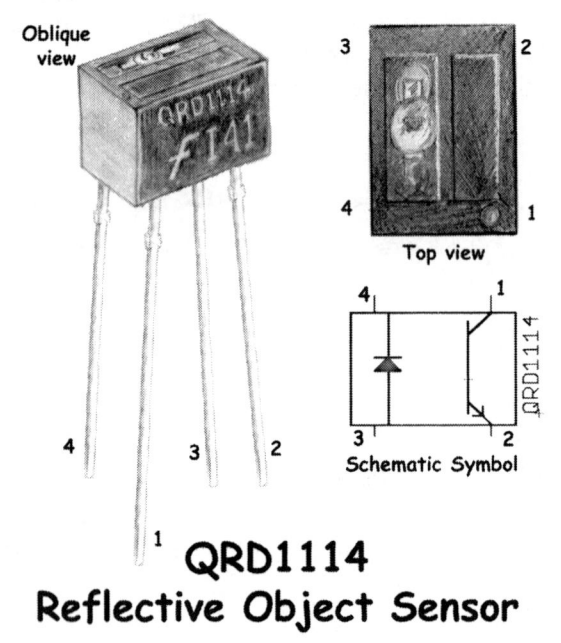

Figure 79: QRD1114 Reflective Object Sensor

Chapter 9: Sensors

We are going to carefully bend the legs on the QRD1114, as shown in Figure 80: Bending the QRD1114 legs, so that we can mount it on the end of the breadboard with it 'looking' to the horizontal. Be very careful when you bend these legs and even more careful when you trim them to fit in the breadboard. The legs will break if bent more than a couple of times so make sure you get it right the first time. It is very easy to lose track of which leg goes where and since you only get one of these in the kit, you don't want to screw this up. Bend the legs that go to the phototransistor straight down to the side and then bend the IR LED legs in the same direction but make the bend a little over 0.1 inch from the bottom. Next trim them so that they will fit in the breadboard with the face of the sensor at the edge of the breadboard (but not overlapping the foamcore base since that edge slides into the ALP box) as shown in the figures below.

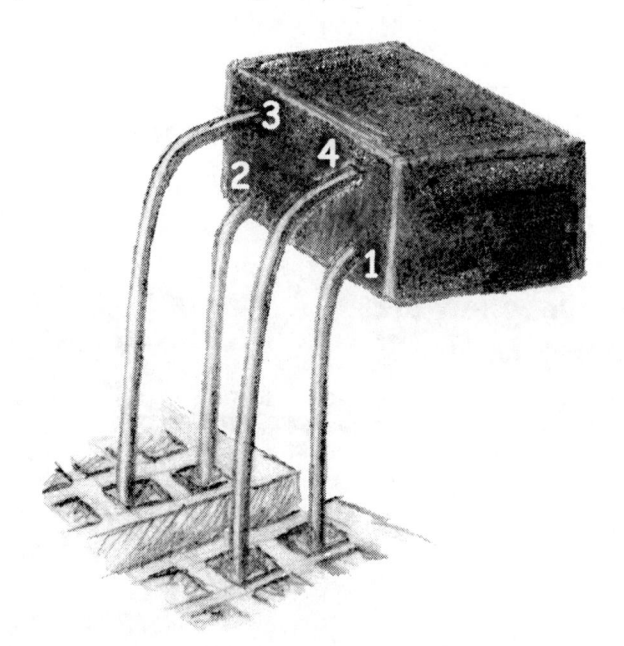

Figure 80: Bending the QRD1114 legs

Figure 81: Object Detector Schematic, Figure 82: IR Object Detector Layout, and Figure 83: IR Object Detector Photo show how to build this circuit.

167

Chapter 9: Sensors

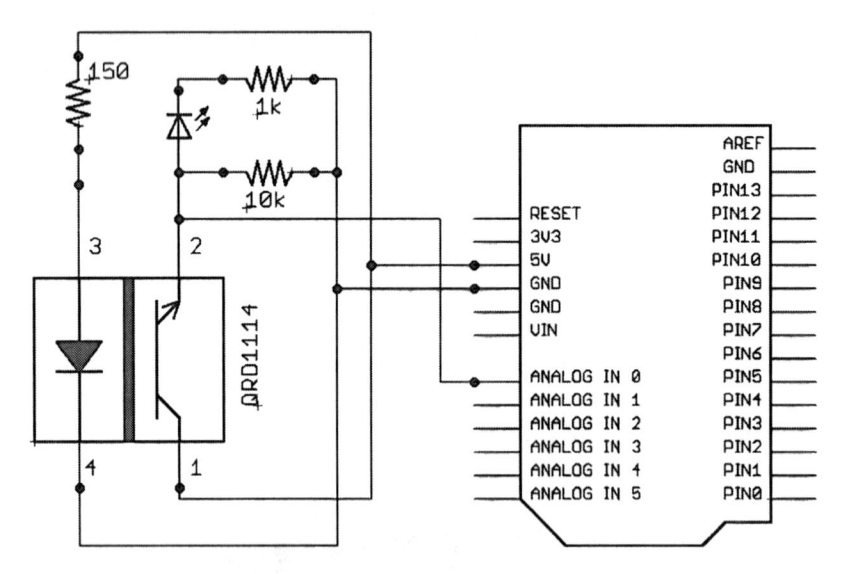

Figure 81: Object Detector Schematic

QRD1114 pin 2 to
10 k Ω to GND and
to LED long led and
to Analog In 0

LED to
1K to GND

QRD1114
legs bent 90°
and offset 0.1"

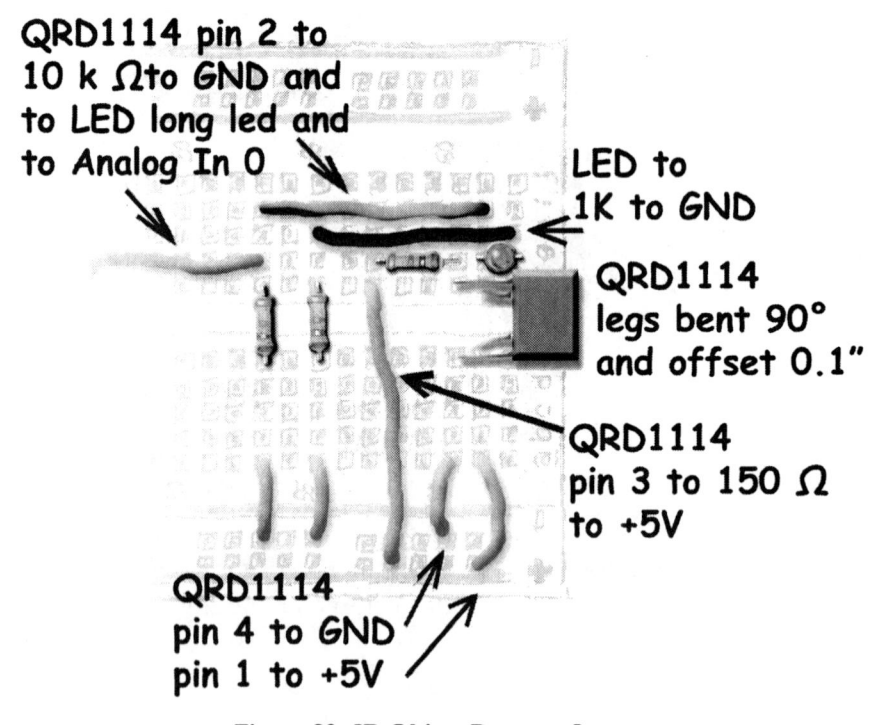

QRD1114
pin 3 to 150 Ω
to +5V

QRD1114
pin 4 to GND
pin 1 to +5V

Figure 82: IR Object Detector Layout

168

Chapter 9: Sensors

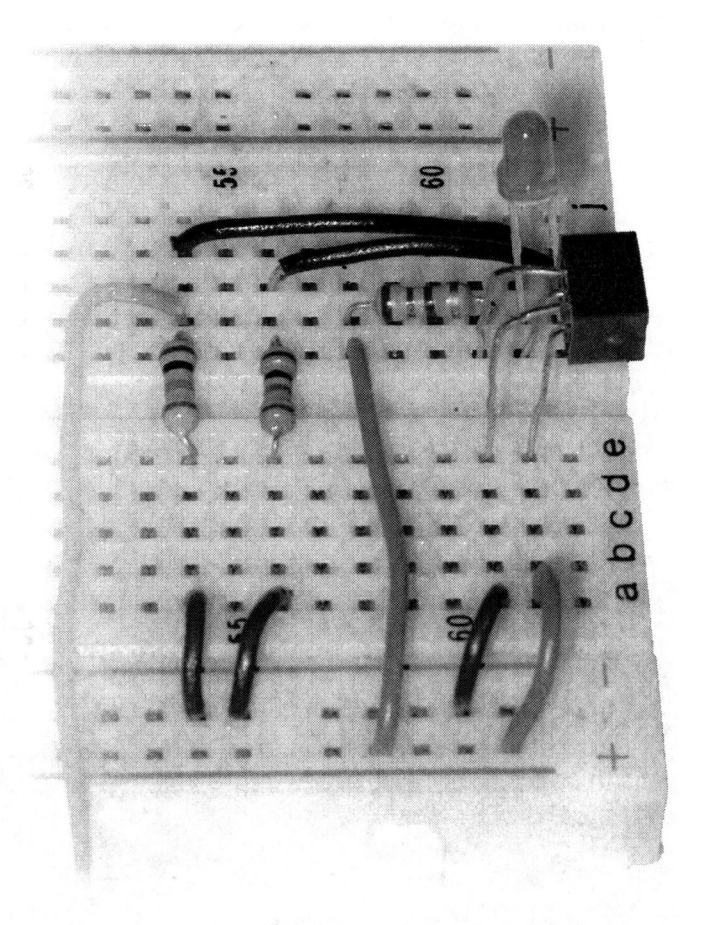

Figure 83: IR Object Detector Photo

As you can see from the schematic, an LED has been inserted into the circuit that will have its brightness controlled by the detector, providing us with a visible cue for the amount of IR being detected. The IR light falling on the phototransistor base provides the current that controls the collector emitter current that flows through the LED (okay, it's a form of magic). You can move your finger back and forth in front of the sensor and the red LED will brighten or dim in response. This is a purely analog response and could, with proper amplification, be used to drive some other analog device such as a motor.

169

Chapter 9: Sensors

We can get a digital measure of this response by putting a 10K Ω resistor from ground to the phototransistor and running a wire from that point to the Arduino ADC in Analog Input pin 0 as shown in the schematic. Let's apply this to a real-world experiment (and by 'real' I mean 'contrived'.)

Tomato Soup Can Counter

Figure 84: Tomato soup can counter prototype

I went a little overboard when I did this experiment, making a foamcore board 'conveyer belt' were I placed six Tomato soup cans (reduced sodium for a

Chapter 9: Sensors

healthier experiment) and dragged them past the sensor to get calibration values. Figure 84: Tomato soup can counter prototype, shows the device in the starting position and after the cans have been conveyed past the sensor.

Construction details: well, I used foamcore board and masking tape and eyeballed everything and so should you – *it's a prototype so expect imperfections*. The 'belt' should be a little wider than a soup can and long enough to hold 6 cans (which is the number I had in my pantry), while the 'base' should be just slightly wider than the 'belt' so you can slide the cans past the QRD1114 as shown in the pictures. The IR detector should almost touch the soup cans as they pass by to get a good reflection.

Tomato Soup Can Counter Software

You have no way to predict the ADC values for the IR returned from a particular object other than it will be within the ADC range between 0 and 1023. I did preliminary tests using a modification of the ReadPot source code in the Arduino IDE substituting the detector for the potentiometer as shown below. The data shown below is taken from a finger moving back and forth in front of the detector. As shown in Figure 85: ADC readings of finger movements, the data varied from a low of 145 to a high of 799 for my finger (your finger will probably be different).

ReadPot Source Code:

```
// ReadPot
// Joe Pardue March 24, 2009
// based on Tom Igoe's example in the
// Arduino Serial.print(data) documentation

void setup()
{
  // begin the serial communication
  Serial.begin(9600);
}

// variable to hold the analog input value
int analogValue = 0;

void loop()
{
```

Chapter 9: Sensors

```
  // read the analog input on pin 0
  analogValue = analogRead(0);

  // print prolog to value
  Serial.print("ADC Reading: ");

  // print as an ASCII-encoded decimal
  Serial.print(analogValue, DEC);
  Serial.println(); // print a newline

  // delay 1 second before the next reading:
  delay(1000);
}
```

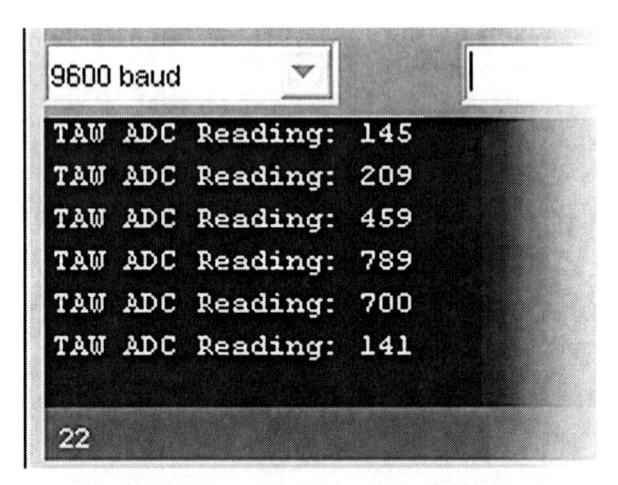

Figure 85: ADC readings of finger movements

Next, I set up the soup can conveyer and slowly moved a can past the sensor while observing the results from the ReadPot software on the Arduino Serial Monitor (slow since the code is only measuring once per second). I saw a range of values above (>600) which I could be certain that a can is present and another range below which I could be sure there is no can present (<300). There is also a sloppy mid-range where I couldn't be sure (301 to 599). (Note: your values won't be the same as mine.) You must make measurements to determine the certain ranges: YES_CAN (for me >600) and NO_CAN (for me <300), then use a true/false variable 'yes' to decide if a can has passed the sensor. There really are many ways to do this kind of thing, but I selected the easy (for me) to understand logic of:

172

Chapter 9: Sensors

If the analogInput is greater than YES_CAN **and** yes is equal 0 **then** set yes to 1.
Else if analogInput is less than NO_CAN **and** yes is equal 1, **then** set yes equal 0 and increment the can count.

Think about this for a moment from the perspective of a can moving by the sensor. Before the can gets there, 'yes' is equal 0 for false and the analogInput value is less than the YES_CAN constant we have predetermined that a can is present. As the can moves into view of the sensor at some point analogValue becomes greater than YES_CAN so we know there is a can present and we set 'yes' equal 1 for true (yup, there is definitely a can present). Then we keep measuring as the can slides past and the analogInput drops to a value less than NO_CAN. We check and see that 'yes' is true so we know there was a can there, but now our analogInput reading says there is no can there so the can must have passed and we set 'yes' to 0 and increment the can count. This will probably be clearer from reading the source code.

I put YES_CAN and NO_CAN in the Tomato_Soup_Can_Counter source code as hard-coded constants (this is not a good programming practice, but it is useful here for keeping things simple). Make sure you change those values to match your measurements.

Tomato_Soup_Can_Counter Source Code:
```
// Tomato_Soup_Can_Counter
// Joe Pardue June 27, 2009

// Sensor level constants
#define NO_CAN 300  // My value may not be yours
#define YES_CAN 600 // My value may not be yours

// 0 if NO_CAN, 1 if YES_CAN
int yes = 0;
// Can count
int count = 0;

// variable for analog input value
int analogValue = 0;

void setup()
```

Chapter 9: Sensors

```
{
  // begin the serial communication
  Serial.begin(9600);
}

void loop()
{
  // read the analog input on pin 0
  analogValue = analogRead(0);

  // print prolog to value
  Serial.print("ADC Reading: ");
  Serial.print(analogValue, DEC);

  // If the analogInput is greater than YES_CAN
  //   and yes is equal 0 set yes to 1.
  // If analogInput is less than NO_CAN
  //   and yes is equal 1, set yes equal 0
  //   and increment count.
  if ((analogValue > YES_CAN) & (yes == 0)){
    yes = 1;
    Serial.print(" - YES_CAN");
  }
  else if ((analogValue < NO_CAN) & (yes == 1)){
    yes = 0;
    Serial.print(" - NO_CAN");
    count++;
  }

  // show the count
  Serial.print(" count = ");
  Serial.print(count, DEC);

  // print a newline
  Serial.println();

  // delay 1 second before the next reading:
  delay(1000);
}
```

For this demonstration I kept the time between ADC readings at one second to avoid being overwhelmed with serial data. When I first tried counting the cans I got some weird can counts and found that I had to make my conveyer belt narrower so that the cans were all pretty close to the sensor when they passed. The

Chapter 9: Sensors

serial output is shown in Figure 86: Can count on Arduino serial monitor. Once you get the system calibrated you should be able to remove the serial output and the one second delay and set a flag (a variable that can be checked to see if an event has occurred) so that you only output the count when it changes.

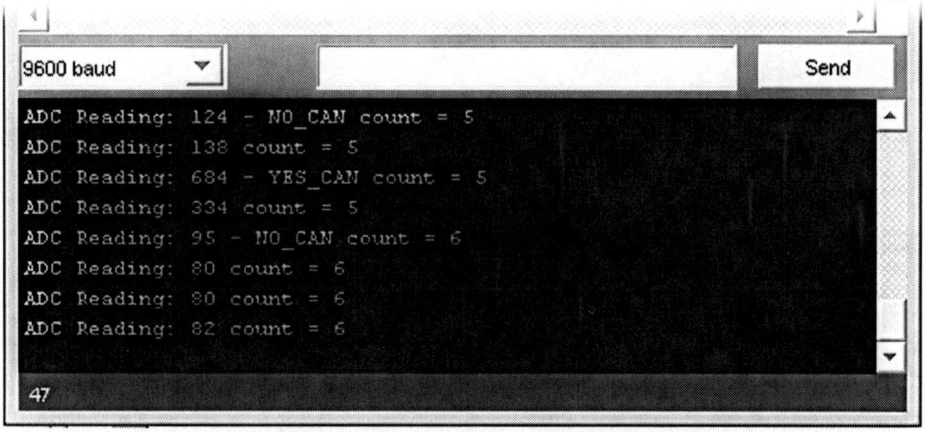

Figure 86: Can count on Arduino serial monitor.

Please note that you should leave the detector circuit in place on the breadboard for the next chapter's circuit.

175

Chapter 10: Simple Motor Speed Control

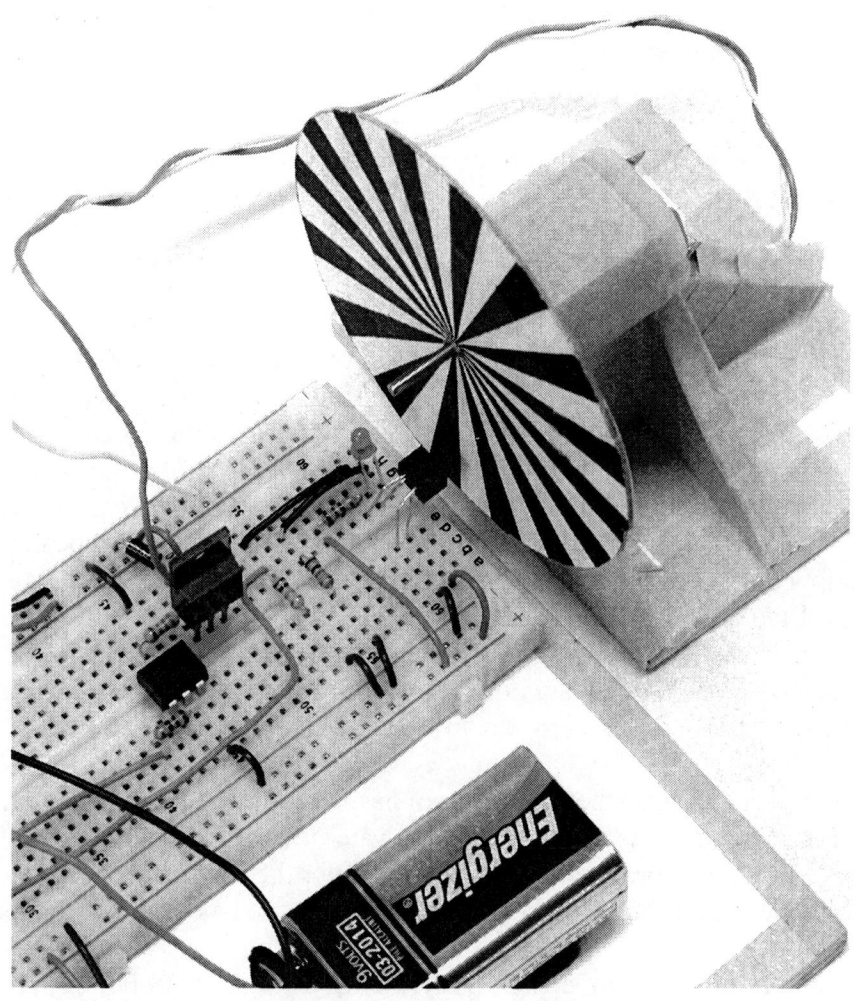

Figure 87: Simple Motor Speed Control

Chapter 10: Simple Motor Speed Control

In this chapter we will look at several seemingly unrelated topics that we will need to understand in order to do simple motor speed control:

1. External interrupts
2. Using the Arduino IDE Serial Monitor to get real data from the PC serial port to the Arduino board.
3. Optical isolation of voltages.

We will use this knowledge to build the circuit shown in Figure 87: Simple Motor Speed Control.

Using external interrupts to detect edges

One of the things you might want a microcontroller to do is perform a service only when a certain external event occurs. For instance, you could put an IR laser beam across a door and have the microcontroller monitor the beam so that when someone passes through the door breaking the beam, the microcontroller turns on the lights (or drops a bucket of water on the intruders head or some such action). If this is the only thing the microcontroller has to do, then it can be dedicated to polling the sensor full time in the loop() function. It would repeatedly run an isItTrippedYet() function that checks the sensor and returns true or false. Polling would drive a person nuts (think of driving kids who are yelling: 'are we there yet', 'are we there yet', 'are we there yet'...) but fortunately microcontrollers don't (yet) care what they are doing, so no need to feel sorry for them if you give them a really boring task. Anyway, it's not like they can retaliate (yet).

But consider the case where the system has lots of other tasks to perform (maybe it is monitoring dozens of doors and the water levels in all the drop buckets). If it is polling each sensor, then someone could enter the room and be beyond the drop zone before the polling tells the system that someone has entered the room.

It is very common to conceptually divide up a microcontroller's work into two groups, one group of routine tasks that can be done any old time and another group of special tasks that must be done immediately when an external event occurs. It is so common, in fact, that most microcontrollers have built-in interrupt peripheral circuitry to accomplish the task of immediate notification.

Chapter 10: Simple Motor Speed Control

This circuitry monitors a pin voltage and when a certain condition happens, such as: 'was high, now low' (falling edge) it generates an interrupt that causes the main program flow to halt, store what it was doing in memory, and then the system runs the function that was assigned to the interrupt. When that function finishes, the system state is restored and the main program runs again from where it was interrupted.

You mainly deal with interrupts in one of two ways. If a simple task is all that is required and the rest of the program doesn't need to know about it, then the interrupt service routine (the function that the AVR calls when the interrupt is tripped) can sneak in, handle it, then sneak back out without the main code ever knowing. If, however, a complex task that takes a lot of time away from the main program needs to be performed, then the interrupt routine should set a flag (change the value of a variable that both it and the main program can see) so that the main program can check that flag as part of the loop() function and deal with the consequences of the interrupt when it gets time. Many novices make the mistake of putting too much code in an interrupt service routine. While 'too much' is relative, you should be okay doing a few quick things like copying pin states to a variable, but you definitely won't be okay to send a query to a terminal program and wait for a response. This is one dog you learn about mostly by being bitten.

The Arduino library function attachInterrupt(interrupt, function, mode) simplifies the chore of setting up and using an external interrupt. The 'interrupt' parameter is either 0 or 1 (for the Arduino digital pin 2 and 3, respectively). The 'function' parameter is the name of the function you want to call when the interrupt occurs (the interrupt service routine). The 'mode' parameter is one of four constants to tell the interrupt when it should be triggered:

      `LOW`: trigger when pin is low
      `CHANGE`: trigger when the pin changes value
      `RISING`: trigger when the pin rises from low to high
      `FALLING`: trigger when the pin falls from high to low

Hopefully, you still have your IR detector setup on the breadboard from the last chapter (see Figure 82: IR Object Detector Layout). All you need to do to the hardware is move the signal wire from the Arduino Analog pin 0 to the Digital

Chapter 10: Simple Motor Speed Control

pin 2 as shown in Figure 88: Edge detection schematic, then run the Edge_Detect_Interrupt software (below). In the former setup, we used the ADC to measure an analog voltage, but this time all we will sense is that the voltage is high enough to represent a digital ON (true) or low enough to represent a digital OFF (false).

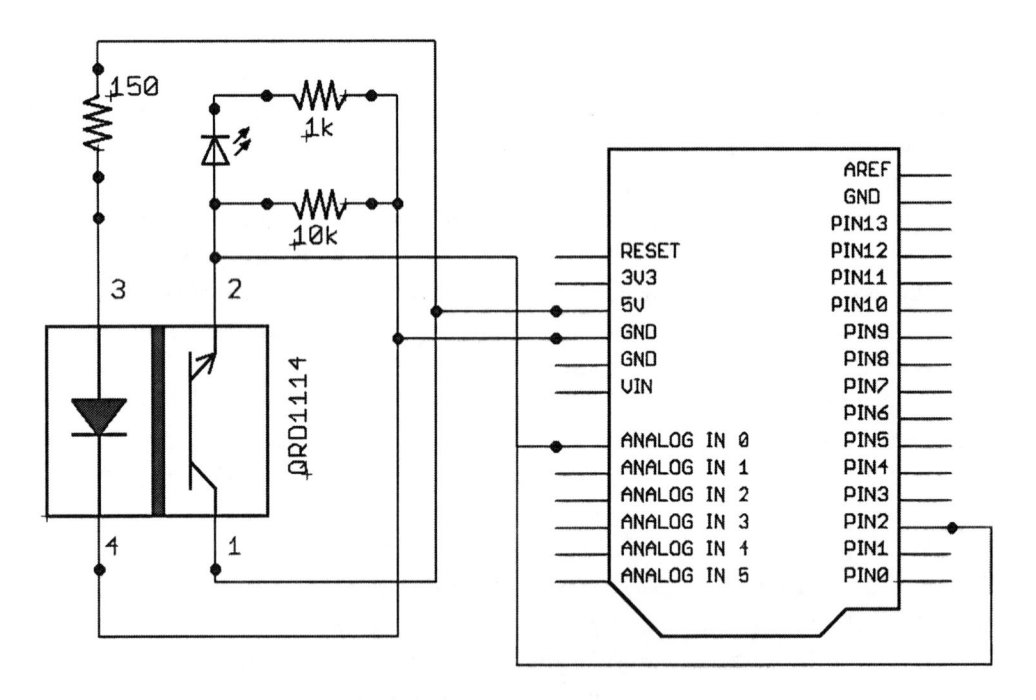

Figure 88: Edge detection schematic

Run the Edge_Detect_Interrupt software and waggle your finger in front of the sensor to get a count like shown in Figure 89: Edge Detect Interrupt counter. Later we will use this concept to detect the passing of black and white stripes on a motor encoder wheel to control the speed of that motor.

Edge Detect Interrupt Software

Edge_Detect_Interrupt Source Code:
```
// Edge_Detect_Interrupt
// Joe Pardue 6/28/09
```

180

Chapter 10: Simple Motor Speed Control

```
volatile int count = 0;

void setup()
{
  // setup the serial port
  Serial.begin(9600);

  // say hello
  Serial.println("Edge Detect Interrupt");

  // attach interrupt 0 (pin 2) to the
  // edgeDetect function
  // run function on falling edge interrupt
  attachInterrupt(0,edgeDetect, FALLING);
}

void loop()
{
// do nothing
}

// on each interrupt
// increment and show the count
void edgeDetect()
{
  count++;
  Serial.print(count);
  Serial.println();
}
```

Figure 89: Edge Detect Interrupt counter

Chapter 10: Simple Motor Speed Control

Optical Isolation of voltage

Have you ever had an EKG and noticed that those wires patched to your chest on either side of your heart go to a machine that is plugged directly into a wall socket capable of providing mains voltage? Now, if you have my kind of worse-case-scenario mind, you may have thought that the particular EKG they were using on you looked a tad dated and maybe a bit rat-infested. You might even have gotten so excited about the prospect of being electrocuted in the doctor's office (how convenient) that you were able to make the EKG go wild with all kinds of crazy beeps while the nurse runs to get the doctor. But never fear, medical devices are designed to prevent lawsuits, which means that as a happy byproduct, the chance of those wires on your chest connecting directly to mains are inversely proportional to the amount a jury would award your estate if it happened.

There are many ways to assure that voltages stay separated, the two main ones are electromagnetic isolation with transformers and optical isolation with LED/phototransistor pairs. We will look at the latter as a way to connect a signal between a microcontroller at 5 volts to a motor at 9 volts so that we can prevent the digital equivalent of a coronary in our micro as a motor jerks the 9-volts around.

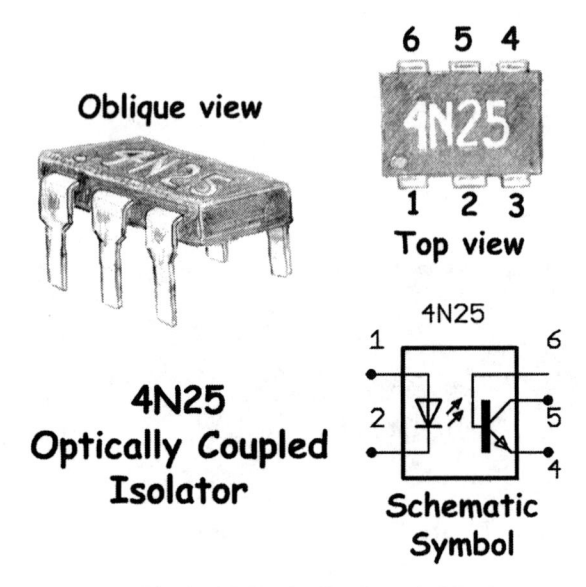

Figure 90: 4N25 Optically Coupled Isolator

Chapter 10: Simple Motor Speed Control

[Aside: note that these DIP packages have a dot or notch to indicate pin 1.]

Figure 90: 4N25 Optically Coupled Isolator provides a drawing and schematic symbol for our optical isolator. You can see that the QRD1114 IR Reflective Object Sensor we looked at in the last chapter and the 4N25 Optically Coupled Isolator that we are about to look at have similar schematic symbols (Figure 91: QRD1114 and 4N25 schematic symbols). Note that the main difference is that the QRD1114 shows a dark bar between the LED (emitter) and the phototransistor (detector) subcomponents. These parts are nearly identical from an electronic perspective. The primary difference is the packaging. The QRD1114 detector is shielded from the emitter and can only 'see' the IR if it is reflected back to the device. The 4N25 is sealed from outside light and the emitter 'shines' directly onto the detector. The QRD1114 detector will pass a current proportional to the reflected IR, thus the signal level is dependant on the external reflective object. The 4N25 responds to the amount of IR coming from the emitter – it can produce a current thorough pins 4 to 5 directly proportional to a current through pins 1 to 2. So if you did something really dumb like connecting a wire from a wall socket to pin 1 on the 4N25, you'd fry the LED, but fortunately none of that voltage would pass through to the device connected to pins 4 or 5. [The datasheet promises ±2500 volts isolation!] This gives us a way to transfer the information in a signal from one circuit to another using light and without having any electrical connection between those circuits. The only connection is in the photons of light.

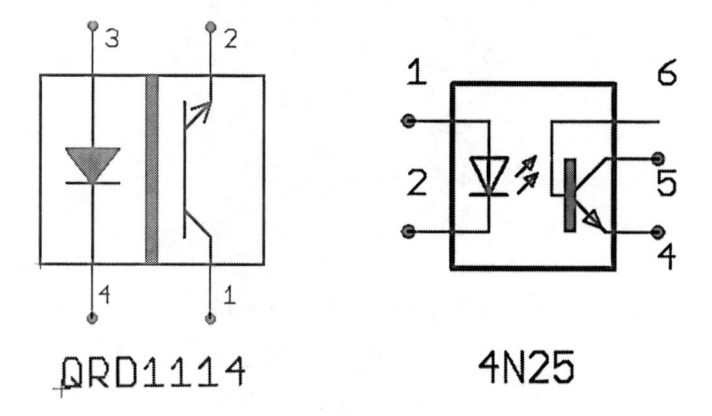

Figure 91: QRD1114 and 4N25 schematic symbols

Chapter 10: Simple Motor Speed Control

Later we will use this device to isolate a PWM signal at one voltage level, 5-volts from our Arduino, and scale it to another voltage level, 9-volts, for our motor driver.

Optical Isolation Component, Schematic, Layout

Our hardware demonstration of these concepts uses a 5-volt signal from the Arduino pin 9 on the emitter side that is converted by the 4N25 to a 9-volt signal on the detector side. Wire this up as shown in the schematic: Figure 93: Optoisolator Test Circuit, and the drawing Figure 92: Optoisolation Test Layout.

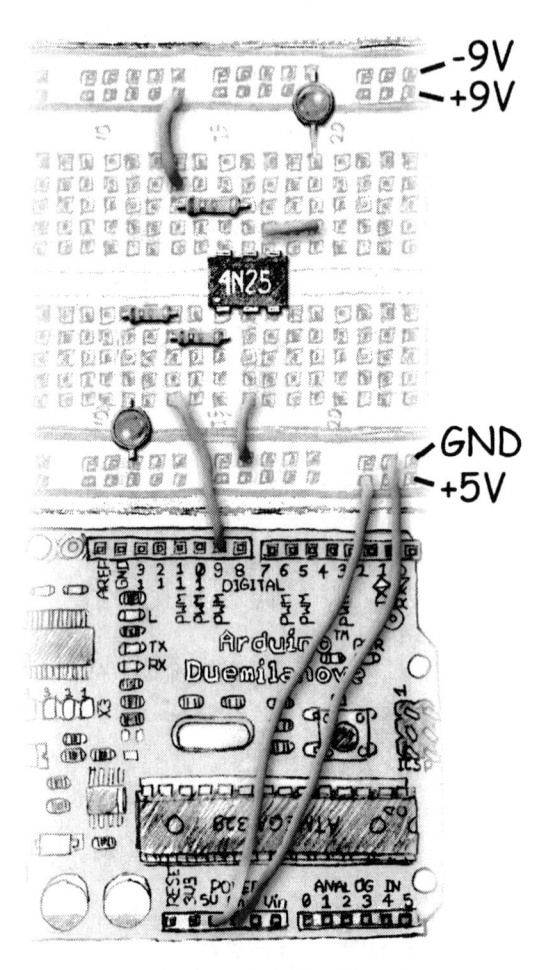

Figure 92: Optoisolation Test Layout

184

Chapter 10: Simple Motor Speed Control

And my apologies for being out of sequence on this, but the 9 volts comes from a battery and the connector as shown in Figure 101: Motor Speed Control Layout. This isn't +9 volts and –9 volts, but the + and – terminals on the battery.

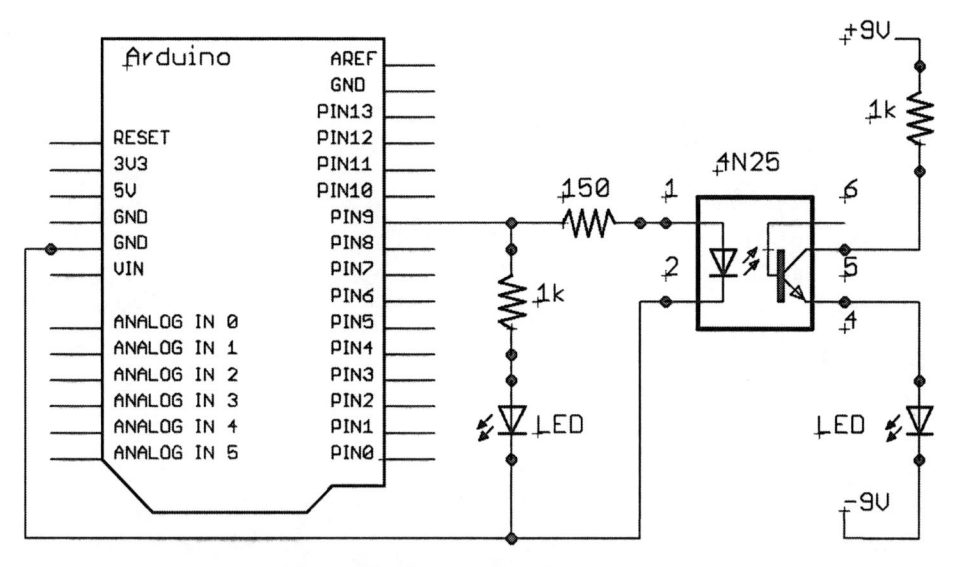

Figure 93: Optoisolator Test Circuit

Optical Isolation Source Code

To test this with software, we will modify the `ASCII_To_Integer` program shown earlier by adding just three lines of code.

PWM_Test Source Code:

```
// PWM_Test 8/1/09 Joe Pardue
// This is the same as ASCII_To_Integer except
// for including the ledpin variable and
// the line adding analogWrite to loop()

int ledpin = 9;    // light connected to digital pin 9

int myInput = 0;
int myNum[6];
int myCount = 0;
int i = 0;
int n = 0;
```

Chapter 10: Simple Motor Speed Control

```
void setup()
{
   Serial.begin(9600);
   Serial.println("PWM Test");
}

void loop()
{
  // get characters until receiving '!'
  while( myInput != '!' ) getNum();

  // convert end-of-number character '!' to 0
  myInput = 0;
  myNum[--myCount] = 0;

  // convert ASCII string to integer
  ATOI();

  // control the brightness of an LED
  analogWrite(ledpin, n);

  // clean up and do it all again
  clearAll();
}

// Put serial characters in a character array
void getNum()
{
  if(Serial.available())
  {
     myInput = Serial.read();
     // put the character in the array
     myNum[myCount++] = myInput;
  }
}

void ATOI()
{
  // algorithm from atoi() in C standard library
  n = 0;
  for(i = 0; myNum[i] >= '0' && myNum[i] <= '9'; ++i)
    n = 10 * n + (myNum[i] - '0');
  // show the number
  Serial.print("You sent: ");
  Serial.println((unsigned int)n,DEC);
}
```

Chapter 10: Simple Motor Speed Control

```
void clearAll()
{
  myCount = 0;
  for(i = 0; i < 6; i++)
  {
    myNum[i] = 0;
  }
  Serial.flush();
}
```

The integer received from the PC, 'n' is converted to a PWM (Pulse Width Modulation) signal on the Arduino pin 9. [I find that values of 200 to 250 work well.] The PWM will cause the LED brightness to be proportional to the input value (within limits). We will discuss PWM in more detail later when we use it to control the speed of a DC motor.

The Arduino pin 9 directly drives the LED on the left (Figure 93: Optoisolator Test Circuit) through a 1k-ohm resistor, and that pin also drives the IR LED in the 4N25 through a 150-ohm resistor. The isolated LED is driven by +9 volts through a 1k-ohm resistor to pin 5 of the 4N25 through the phototransistor out pin 4 to the LED.

Chapter 10: Simple Motor Speed Control

How our DC motor works

Rotor: Wire loop electromagnet reverses north/south polarity each rotation

Stator: Permanent magnet south pole

Stator: Permanent magnet north pole

−9V +9V

Commutator: Reverses current on each rotation

Rotation around axis

Sleeve attaches to motor wire loop and slides inside brush

Axle

Gap between sleeves

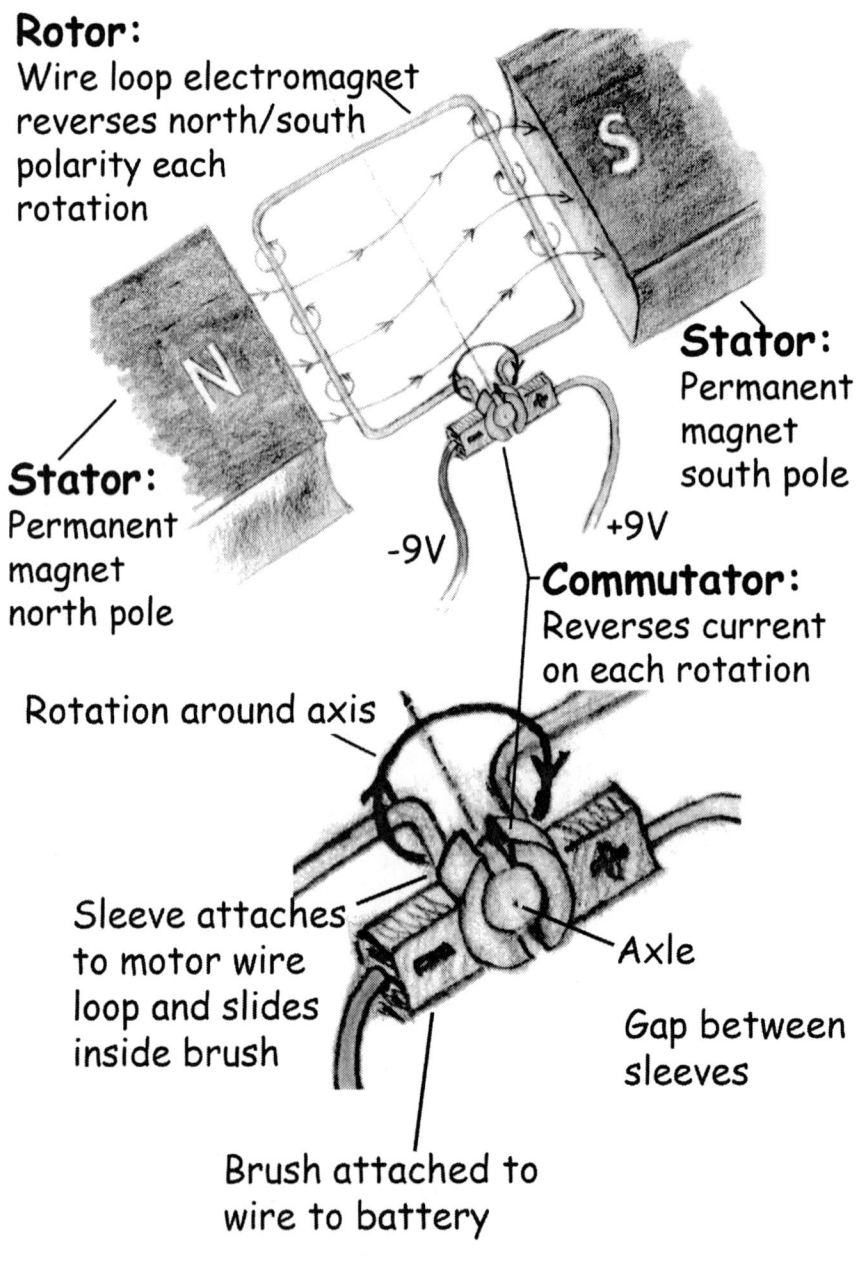

Brush attached to wire to battery

Figure 94: DC Motor Principles

Chapter 10: Simple Motor Speed Control

In Figure 94: DC Motor Principles we see a simplified drawing showing how a DC motor runs. There are three main components:

1. Stator
2. Rotor
3. Commutator

The stator, as shown here is a pair of permanent magnets that don't move (static). The rotor is a loop of copper wire forming an electromagnet that - wait for it... rotates. And the commutator is a clever little mechanical device that takes the ends of the loop, flattens them on a cylindrical sleeve over the axle so that they have a gap between them (preventing the ends from short circuiting) – one end of the loop wire on one side of the sleeve and the other end on the other sleeve so that they can slide under the brushes effectively reversing the voltage each time the rotor rotates. In the figure, the electromagnet is shown aligned with the permanent magnets, but the electromagnetic field is aligned in opposite directions causing the distortion in the permanent magnetic field. This means that the part of the loop to the left is the electromagnetic north and attracted to the south pole of the permanent magnet on the right, and the right side of the loop is south and attracted to the north magnet. So the magnetic attraction/repulsion causes the loop to turn in the clockwise direction. But look what happens to the loop end sleeves on the commutator as the loop turns. The loop end that was touching the negative electric brush rotates away and comes in contact with the positive electric brush while the end that was touching the positive brush now contacts the negative brush. The current reverses in the loop causing the electro magnetism to reverse; the side that was attracted to the south magnetic pole is now attracted to the north magnetic pole. The rotation continues on around clockwise pulled by the magnetic forces until the loops nearly attain their desired goal when the commutator again causes the current and thus the magnetic attraction to reverse keeping the loop spinning about the axle. For some reason when I was writing this I began to wonder if maybe youthful romance doesn't have some sort of commutator that causes attraction to opposites until they get near then find themselves repulsed and attracted to the other opposite and so on until their axle wears out (followed by marriage, kids, debt, attempts to lube the commutator, overheating, short-circuits, nursing home...).

Chapter 10: Simple Motor Speed Control

And while this description does cover the principles of DC motors (and romance) it oversimplifies both. The commutator as drawn will short circuit once each turn when the gaps are under the brushes, so the actual motor we are using (Figure 95: DC motor dismantled) has a three point commutator with three iron posts for the winding which not only prevents the short, but makes for better sequencing of the magnetic attractions/repulsions. (I'll forbear any more romantic metaphors, [though there is often a third party involved...]).

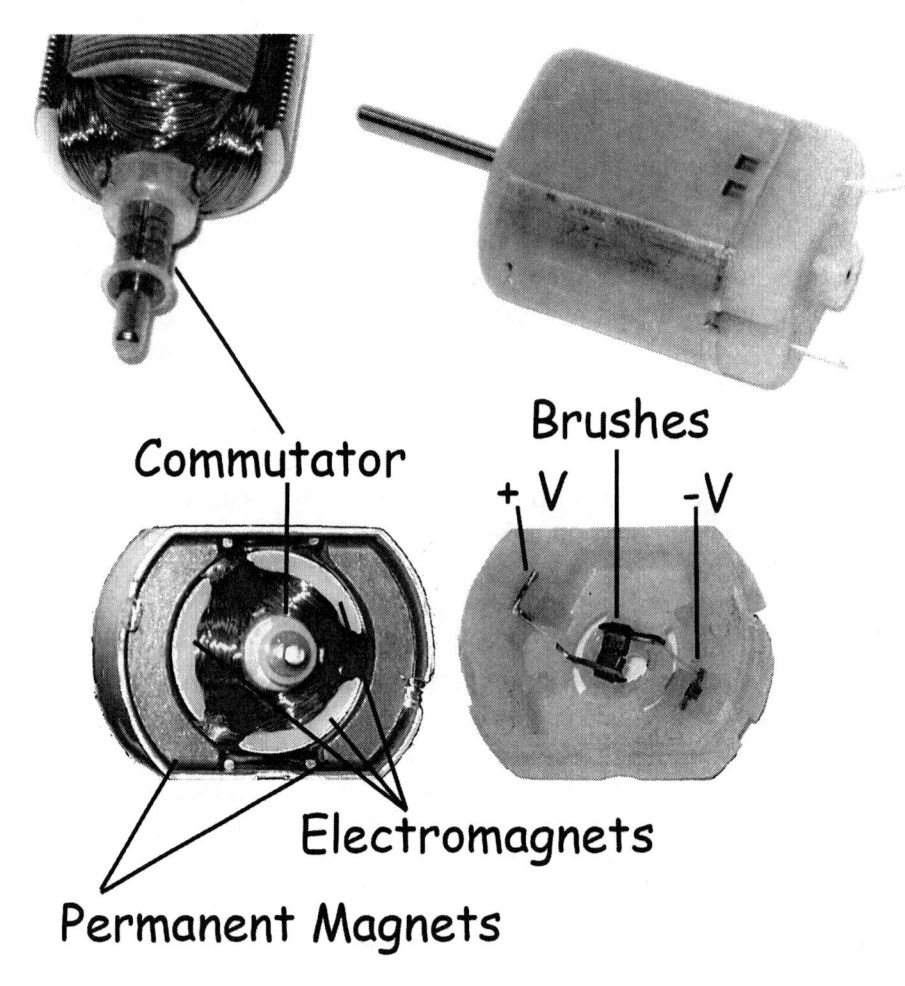

Figure 95: DC motor dismantled

Chapter 10: Simple Motor Speed Control

Diode to suppress voltage spikes

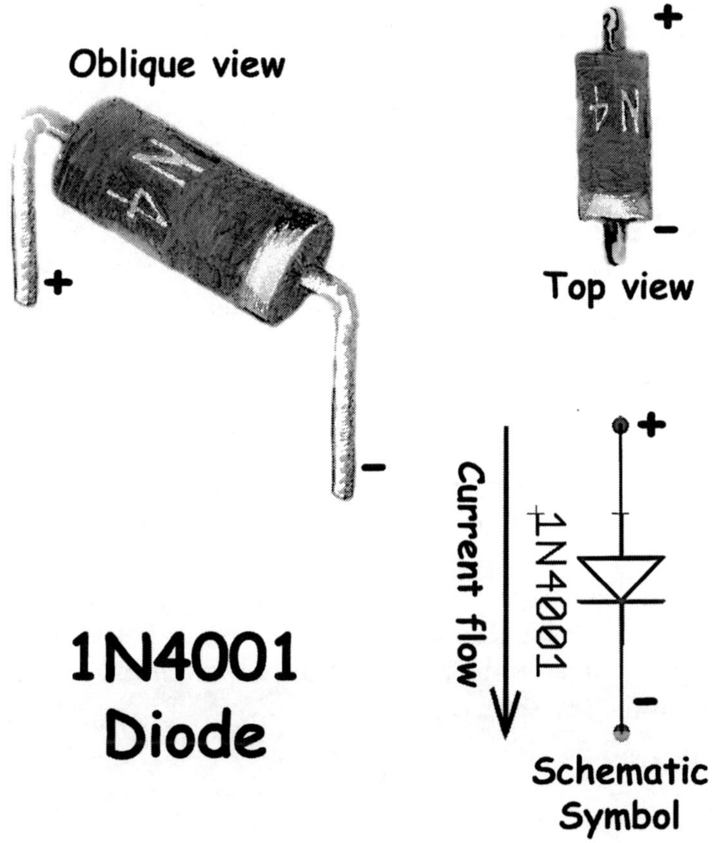

Figure 96: Diode

The process involved in making the motor turn also causes the current to reverse in the copper windings every turn. One notable characteristic of coils of wire (like in the motor windings) is that once the current has started flowing, it doesn't want to stop. If you try to stop the current by cutting the wire, the current will 'pile up' on one side of the cut creating a high voltage that can drive the current through the air across the cut as a spark. (I'm sure there is a romantic metaphor in here somewhere about being cut off causing a rapid rise in voltage with associated sparks flying.) In the motor, the current must not only stop, it must reverse directions for each turn of the motor. This stop and reversal process generates high voltage spikes for each revolution of the motor and while an isolated motor

Chapter 10: Simple Motor Speed Control

can handle this with no problems, the voltage spikes can wreak havoc on other devices that share the same power supply. The diode shown in Figure 100: Motor Speed Control Schematic, acts like a one-way valve so that when the current is flowing in the proper direction the valve is off, but when the current backs up, the valve opens to let it drain off the reverse surge.

Powering the motor

The motor in the Arduino Projects Kit is designed to run from 6 to 15 volts (nominal 12 volts, but 9 volts is fine for our purposes.) and about 150 milliamps. There is a lot of slop in these figures and you can get it to turn with lower voltage or current and it will spin happily at higher voltages or currents, but below certain values it won't turn and above certain values it will heat up and something will break. We will use our battery as a constant (more or less) voltage source, and control the motor speed by pulsing the current with a transistor (PWM).

And FYI, every explanation I've read about how transistors work has been either too simple or too complex, so let's just accept that a tiny current on the base pin controls a much larger current between the collector and emitter pins by magic (Figure 99: Power Transistor).

Chapter 10: Simple Motor Speed Control

Using PWM to control the motor speed

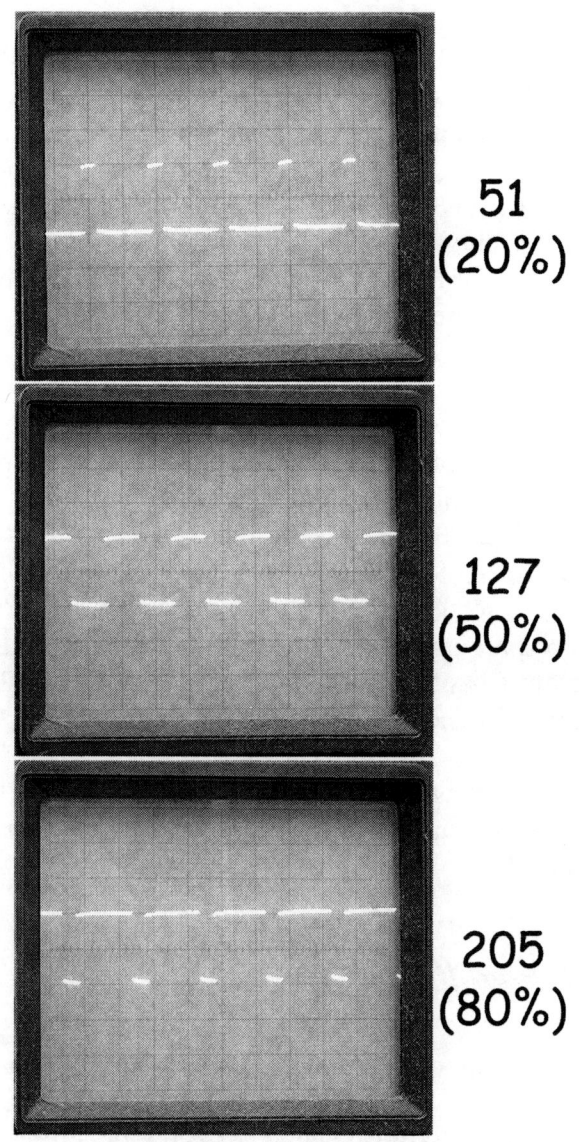

Figure 97: PWMs on oscilloscope

Chapter 10: Simple Motor Speed Control

We will use a Pulse Width Modulation (PWM) signal transmitted from the Arduino through our optoisolator to the base of our TIP115 transistor to make or break the connection to our 9-volt battery.

The Arduino analogWrite() function produces a PWM signal with a frequency of about 490Hz (on/off periods per second). During each of these periods the signal can be turned on for a part of the period and off for a part of the period. The on/off time is called the duty cycle. It can vary from 0 (fully off) to 255 (fully on) and increments in between such as 127, which sets it 'on' half the time and 'off' half the time (50% duty cycle). As you can see from Figure 97: PWMs on oscilloscope, a value of 51 sets a 20% on time for each of the cycles, and a value of 205 sets an 80% on time for each cycle.

The motor will run slower at a low duty cycle and faster at a high duty cycle, but the relative speeds are not directly proportional to the duty cycle. You need a minimum duty cycle to provide enough energy to get the motor going - in my case sending analogWrite() a value below 25 wouldn't make it run.

The point is that you can't know the motor speed just from the duty-cycle you are generating. You have to actually measure the speed and then adjust the duty-cycle to fit the speed you require. (We'll do this in a moment reusing the IR Detector Interrupt code from Chapter 9). But first build the circuit shown in Figure 100: Motor Speed Control Schematic, and Figure 101: Motor Speed Control Layout. Test this circuit with the Arduino Fade example code (it is part of the Arduino IDE example code and was discussed Chapter 5). If you hold the motor, you should feel it speeding up and slowing down to the same timing as the LED brightened and faded.

Building the breadboard circuit

This is the most complicated circuit we will be building using the Arduino Projects Kit and, frankly, the chances of building the full circuit and writing the code from scratch and having it work correctly first time are almost nil. You should think of this as built from hardware/software sub-components that we've done before. First, make sure the IR detector is working properly (built and tested in Chapter 9), then make sure the optoisolator is doing what it should be doing

Chapter 10: Simple Motor Speed Control

(built and tested earlier in this Chapter). Next we add the TIP115 to the optoisolator circuit in place of the LED and test that the motor speed varies like the LED brightness varied. Only after you are sure that each part is working properly should you put the encoder wheel on the motor and try to use it to control the motor speed. This is a breadboard and something **will** go wrong, so be prepared to take small steps and when something goes wrong, be willing to back up and verify each part of the whole.

Please note that the photograph in Figure 87: Simple Motor Speed Control, shows the power on the opposite end of the breadboard than what is shown in the drawing in Figure 101: Motor Speed Control Layout. I did this to simplify the circuit by showing it isolated and not mixed up with the IR detector circuitry, but it shouldn't matter where you put either as long as the QRD1114 is sticking out over the end of the board close to the encoder wheel.

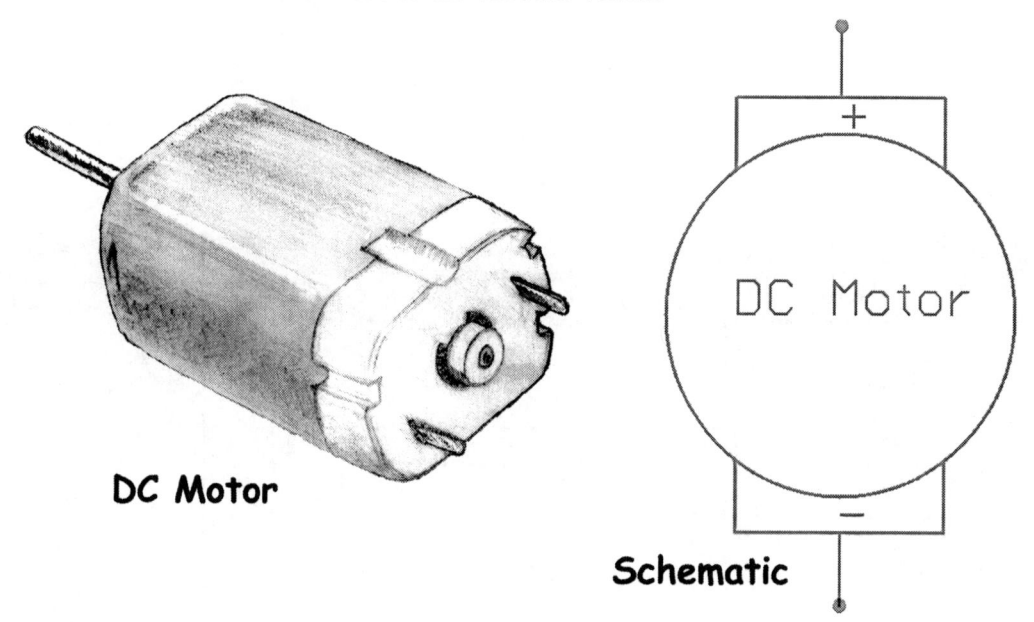

Figure 98: DC Motor

Chapter 10: Simple Motor Speed Control

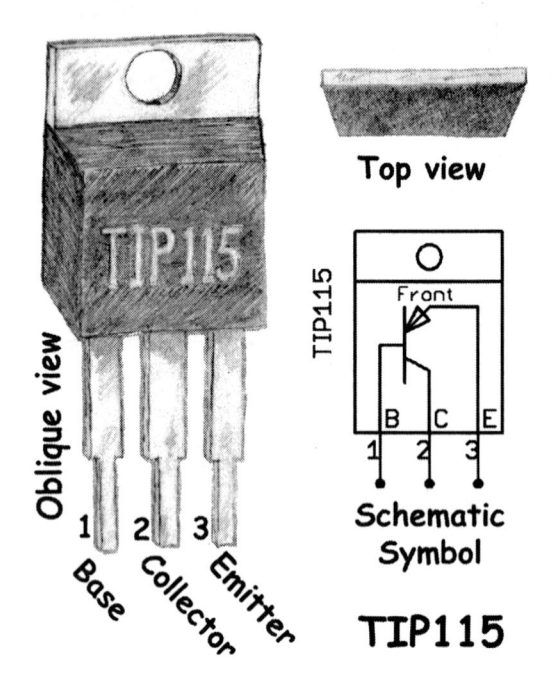

Figure 99: Power Transistor

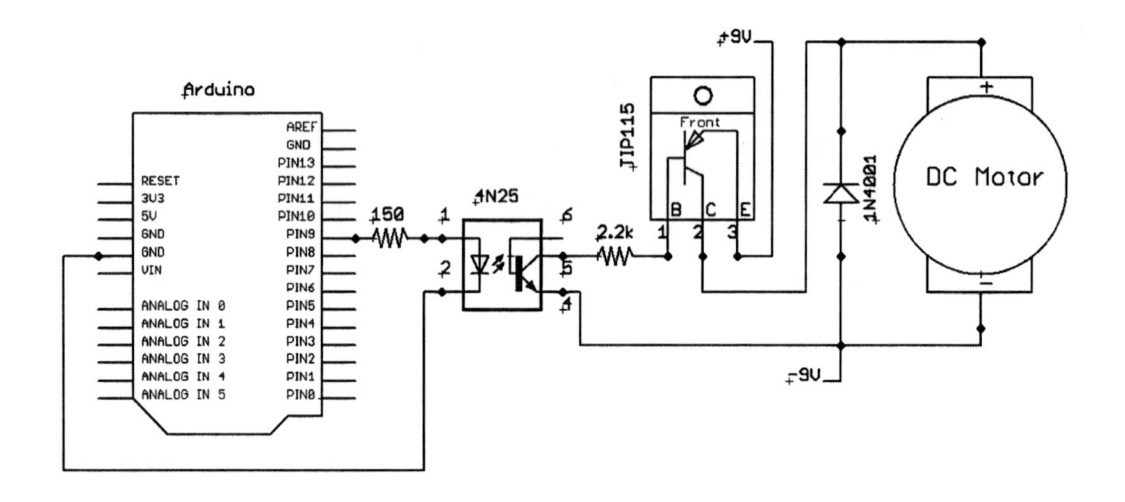

Figure 100: Motor Speed Control Schematic

Chapter 10: Simple Motor Speed Control

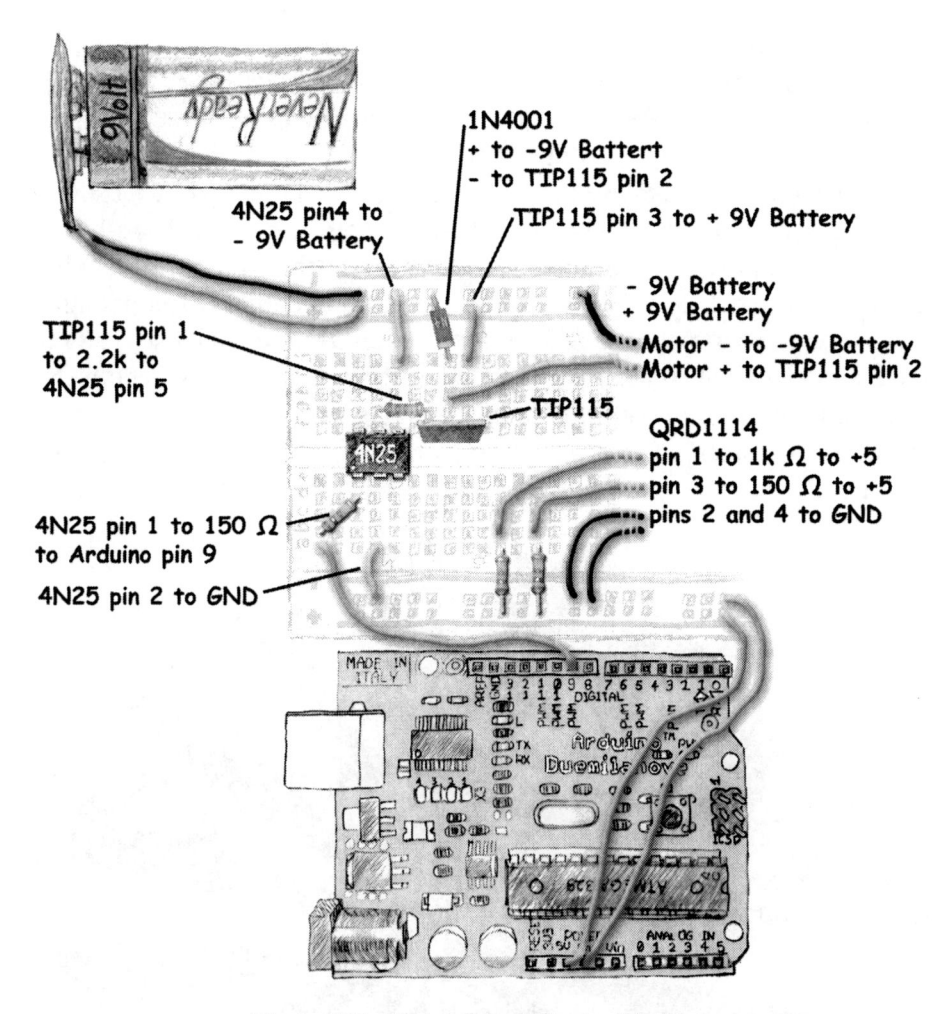

Figure 101: Motor Speed Control Layout

Aside: Be sure to keep the +5V separate from the +9V as shown above.

Using an encoder wheel to measure the motor speed

We will reuse the IR reflective sensor circuit and the interrupt from Chapter 9 to count the passing of the stripes on the encoder wheel shown in Figure 102: Encoder Wheel. You might get by with cutting the encoder wheel out of the book and pasting it on some cereal box pasteboard, but I'm not sure about the actual reflectivity of IR from the surface of this book paper, so I'll recommend that you download the pdf file of the image and print it on plain paper with an inkjet

Chapter 10: Simple Motor Speed Control

printer (you can get the encoder file from my website), then, for even better IR reflectivity, darken the black stripes where they will be in front of the QRD1114 using a Sharpie © pen. Paste the disk on a piece of cardboard (a cereal box should do fine, but you may need to put some weights on the disk while it is drying to keep it flat.) When it is dry, make a hole for the motor axle by using an Xacto knife or a scalpel or whatever very sharp pointed thing you have handy, slice a few 1/8" cuts in the form of and asterisk * at the center point of the wheel. In my case, I could slip the wheel on the motor axle and there was enough pressure provided by the cardboard to hold it in place, but if yours is loose, you might want to add a touch of glue – after you've made certain that the disk is both flat and at a 90° angle to the axle. You might let the glue get tacky, then run the motor while it finishes drying so the centrifugal force will align the disk properly. Play with it since a little wobble won't hurt, but a lot of wobble may make the QRD1114 readings unreliable.

Figure 102: Encoder Wheel

Chapter 10: Simple Motor Speed Control

Simple motor speed control with digital feedback

The program Simple_Motor_Speed_Control uses many of the principles we've been discussing. To set the speed, from the Arduino IDE serial monitor you enter a number followed by an '!'. This number will be compared to the count from the encoder wheel spinning in front of the IR detector. If the actual count is lower than the input value then the value being sent to the PWM by analogWrite(value) will be incremented by the amount in the constant ADJUST (5 in this case). And if the count is greater than the input number the value will be decremented. You can find the maximum and minimum input values by experimenting. I noted that values of less than 125 caused the motor to stop and values greater than 1050 maxed out the PWM value. Figure 103: Program Serial I/O, shows that entering a value of 200 for the 'Input' when the 'Count' is 596 causes the 'Speed' to decrease by 5 each second. When the count is close to the input the speed will increase and decrease each second to keep the count close to the input. Even though the hardware and software are 'simple' it serves to show the basic principles involved for one method of motor speed control.

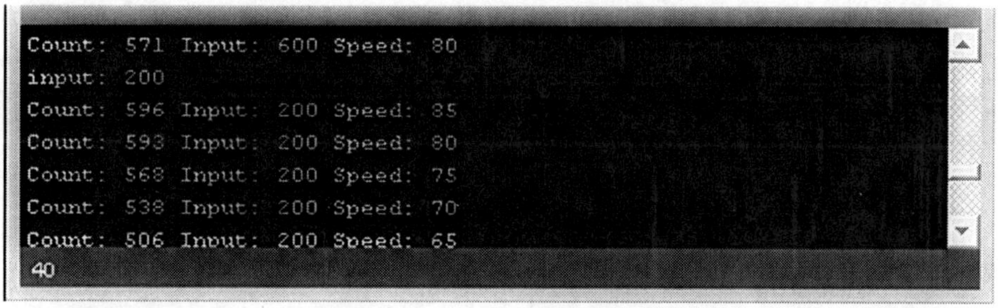

Figure 103: Program Serial I/O

Simple_Motor_Speed_Control Source Code:

```
// Simple_Motor_Speed_Control 8/13/09 Joe Pardue

#define ADJUST 5 // speed +or-

// variable to keep PWM value
int value = 0;
// pin for motor PWM signal
int motorpin = 9;

// variables for serial input
```

199

Chapter 10: Simple Motor Speed Control

```
int myInput = 0;
int myNum[6];
int myCount = 0;

// always declare interrupt variables
// as volatile
volatile int count = 0;

// serial input converted to integer
int input = 0;

// value for PWM
int speed = 0;

// time keeping
long oldTime = 0;
long newTime = 0;

void setup()
{
   Serial.begin(9600);
   Serial.println("Simple_Motor_Speed_Control");

  // attach interrupt 0 (pin 2) to the
  // edgeDetect function
  // run function on falling edge interrupt
  attachInterrupt(0,edgeDetect, FALLING);

  oldTime = millis();
}

void loop()
{
  newTime = millis();
  if(newTime > (oldTime + 1000))
  {
    oldTime = newTime;
    Serial.print("Count: ");
    Serial.print(count);
    Serial.print(" Input: ");
    Serial.print(input);
    Serial.print(" Speed: ");
    Serial.print(speed);
    Serial.println();

    if( (speed >= 0)&&(speed<=255) )
```

Chapter 10: Simple Motor Speed Control

```
    {
      if(count < input)
      {
        if (speed != 255)
        {
          speed += ADJUST;
        }
      }
      else
      {
        if (speed != 0)
        {
          speed -= ADJUST;
        }
      }
      analogWrite(motorpin, speed);
    }
    else (speed = 0);

    count = 0;
  }

  getNum();
  if(myInput == '!')
  {
    // convert end-of-number character '!' to 0
    myInput = 0;
    myNum[--myCount] = 0;

    // convert ASCII string to integer
    input = ATOI();

    // map the count number to the PWM value
    Serial.print("input: ");
    Serial.println(input,DEC);

    // clean up and do it all again
    clearAll();
  }
}

// Put serial characters in a character array
void getNum()
{
  if(Serial.available())
  {
```

201

Chapter 10: Simple Motor Speed Control

```
      myInput = Serial.read();
      // put the character in the array
      myNum[myCount++] = myInput;
   }
}

int ATOI()
{
   // algorithm from atoi() in C standard library
   int i = 0;
   int n = 0;
   for(i = 0; myNum[i] >= '0' && myNum[i] <= '9'; ++i)
     n = 10 * n + (myNum[i] - '0');

   return(n);
}

void clearAll()
{
   int i;

   myCount = 0;
   for(i = 0; i < 6; i++)
   {
     myNum[i] = 0;
   }
   Serial.flush();
}

// On each IR detector interrupt
// increment the count
void edgeDetect()
{
   count++;
}
```

Chapter 10: Simple Motor Speed Control

Now what?

Well that's it for this Arduino Workshop book. You have got a good start on playing with microcontrollers and you might be surprised to learn that you've also gotten a good start in learning the C programming language – ***the*** professional microcontroller programming language (remember what I said about IMHO Chapter 1?) If you didn't find this book too frustrating and want to dig deeper into microcontrollers then I suggest you look at www.smileymicros.com. There is just no telling what sort of surprising new things you might find there.

I hope you had as much fun playing with the Arduino as I had. See you later?

Appendix 1: ASCII Table

Table 2: ASCII Table

```
Char   Dec   Hex  | Char  Dec   Hex  | Char  Dec   Hex  | Char  Dec   Hex
------------------------------------------------------------------------
(nul)   0    0x00 | (sp)   32   0x20 |  @     64   0x40 |  `     96   0x60
(soh)   1    0x01 |  !     33   0x21 |  A     65   0x41 |  a     97   0x61
(stx)   2    0x02 |  "     34   0x22 |  B     66   0x42 |  b     98   0x62
(etx)   3    0x03 |  #     35   0x23 |  C     67   0x43 |  c     99   0x63
(eot)   4    0x04 |  $     36   0x24 |  D     68   0x44 |  d    100   0x64
(enq)   5    0x05 |  %     37   0x25 |  E     69   0x45 |  e    101   0x65
(ack)   6    0x06 |  &     38   0x26 |  F     70   0x46 |  f    102   0x66
(bel)   7    0x07 |  '     39   0x27 |  G     71   0x47 |  g    103   0x67
(bs)    8    0x08 |  (     40   0x28 |  H     72   0x48 |  h    104   0x68
(ht)    9    0x09 |  )     41   0x29 |  I     73   0x49 |  i    105   0x69
(nl)   10    0x0a |  *     42   0x2a |  J     74   0x4a |  j    106   0x6a
(vt)   11    0x0b |  +     43   0x2b |  K     75   0x4b |  k    107   0x6b
(np)   12    0x0c |  ,     44   0x2c |  L     76   0x4c |  l    108   0x6c
(cr)   13    0x0d |  -     45   0x2d |  M     77   0x4d |  m    109   0x6d
(so)   14    0x0e |  .     46   0x2e |  N     78   0x4e |  n    110   0x6e
(si)   15    0x0f |  /     47   0x2f |  O     79   0x4f |  o    111   0x6f
(dle)  16    0x10 |  0     48   0x30 |  P     80   0x50 |  p    112   0x70
(dc1)  17    0x11 |  1     49   0x31 |  Q     81   0x51 |  q    113   0x71
(dc2)  18    0x12 |  2     50   0x32 |  R     82   0x52 |  r    114   0x72
(dc3)  19    0x13 |  3     51   0x33 |  S     83   0x53 |  s    115   0x73
(dc4)  20    0x14 |  4     52   0x34 |  T     84   0x54 |  t    116   0x74
(nak)  21    0x15 |  5     53   0x35 |  U     85   0x55 |  u    117   0x75
(syn)  22    0x16 |  6     54   0x36 |  V     86   0x56 |  v    118   0x76
(etb)  23    0x17 |  7     55   0x37 |  W     87   0x57 |  w    119   0x77
(can)  24    0x18 |  8     56   0x38 |  X     88   0x58 |  x    120   0x78
(em)   25    0x19 |  9     57   0x39 |  Y     89   0x59 |  y    121   0x79
(sub)  26    0x1a |  :     58   0x3a |  Z     90   0x5a |  z    122   0x7a
(esc)  27    0x1b |  ;     59   0x3b |  [     91   0x5b |  {    123   0x7b
(fs)   28    0x1c |  <     60   0x3c |  \     92   0x5c |  |    124   0x7c
(gs)   29    0x1d |  =     61   0x3d |  ]     93   0x5d |  }    125   0x7d
(rs)   30    0x1e |  >     62   0x3e |  ^     94   0x5e |  ~    126   0x7e
(us)   31    0x1f |  ?     63   0x3f |  _     95   0x5f | (del) 127   0x7f
```

```
Name   Description         C Escape Sequence
nul    null byte                \0
bel    bell character           \a
bs     backspace                \b
ht     horizontal tab           \t
np     formfeed                 \f
nl     newline                  \n
cr     carriage return          \r
vt     vertical tab             \v
```

Appendix 2: Decimal, Hexadecimal, and Binary

Table 3: Decimal, Hexadecimal, and Binary Conversion

Dec	Hex	Bin	Dec	Hex	Bin	Dec	Hex	Bin	Dec	Hex	Bin
0	0	00000000	64	40	01000000	128	80	10000000	192	c0	11000000
1	1	00000001	65	41	01000001	129	81	10000001	193	c1	11000001
2	2	00000010	66	42	01000010	130	82	10000010	194	c2	11000010
3	3	00000011	67	43	01000011	131	83	10000011	195	c3	11000011
4	4	00000100	68	44	01000100	132	84	10000100	196	c4	11000100
5	5	00000101	69	45	01000101	133	85	10000101	197	c5	11000101
6	6	00000110	70	46	01000110	134	86	10000110	198	c6	11000110
7	7	00000111	71	47	01000111	135	87	10000111	199	c7	11000111
8	8	00001000	72	48	01001000	136	88	10001000	200	c8	11001000
9	9	00001001	73	49	01001001	137	89	10001001	201	c9	11001001
10	a	00001010	74	4a	01001010	138	8a	10001010	202	ca	11001010
11	b	00001011	75	4b	01001011	139	8b	10001011	203	cb	11001011
12	c	00001100	76	4c	01001100	140	8c	10001100	204	cc	11001100
13	d	00001101	77	4d	01001101	141	8d	10001101	205	cd	11001101
14	e	00001110	78	4e	01001110	142	8e	10001110	206	ce	11001110
15	f	00001111	79	4f	01001111	143	8f	10001111	207	cf	11001111
16	10	00010000	80	50	01010000	144	90	10010000	208	d0	11010000
17	11	00010001	81	51	01010001	145	91	10010001	209	d1	11010001
18	12	00010010	82	52	01010010	146	92	10010010	210	d2	11010010
19	13	00010011	83	53	01010011	147	93	10010011	211	d3	11010011
20	14	00010100	84	54	01010100	148	94	10010100	212	d4	11010100
21	15	00010101	85	55	01010101	149	95	10010101	213	d5	11010101
22	16	00010110	86	56	01010110	150	96	10010110	214	d6	11010110
23	17	00010111	87	57	01010111	151	97	10010111	215	d7	11010111
24	18	00011000	88	58	01011000	152	98	10011000	216	d8	11011000
25	19	00011001	89	59	01011001	153	99	10011001	217	d9	11011001
26	1a	00011010	90	5a	01011010	154	9a	10011010	218	da	11011010
27	1b	00011011	91	5b	01011011	155	9b	10011011	219	db	11011011
28	1c	00011100	92	5c	01011100	156	9c	10011100	220	dc	11011100
29	1d	00011101	93	5d	01011101	157	9d	10011101	221	dd	11011101
30	1e	00011110	94	5e	01011110	158	9e	10011110	222	de	11011110
31	1f	00011111	95	5f	01011111	159	9f	10011111	223	df	11011111
32	20	00100000	96	60	01100000	160	a0	10100000	224	e0	11100000
33	21	00100001	97	61	01100001	161	a1	10100001	225	e1	11100001
34	22	00100010	98	62	01100010	162	a2	10100010	226	e2	11100010
35	23	00100011	99	63	01100011	163	a3	10100011	227	e3	11100011
36	24	00100100	100	64	01100100	164	a4	10100100	228	e4	11100100
37	25	00100101	101	65	01100101	165	a5	10100101	229	e5	11100101
38	26	00100110	102	66	01100110	166	a6	10100110	230	e6	11100110
39	27	00100111	103	67	01100111	167	a7	10100111	231	e7	11100111
40	28	00101000	104	68	01101000	168	a8	10101000	232	e8	11101000
41	29	00101001	105	69	01101001	169	a9	10101001	233	e9	11101001
42	2a	00101010	106	6a	01101010	170	aa	10101010	234	ea	11101010
43	2b	00101011	107	6b	01101011	171	ab	10101011	235	eb	11101011

Appendix 2: Decimal, Hexadecimal, and Binary

```
44 2c 00101100    108 6c 01101100    172 ac 10101100    236 ec 11101100
45 2d 00101101    109 6d 01101101    173 ad 10101101    237 ed 11101101
46 2e 00101110    110 6e 01101110    174 ae 10101110    238 ee 11101110
47 2f 00101111    111 6f 01101111    175 af 10101111    239 ef 11101111
48 30 00110000    112 70 01110000    176 b0 10110000    240 f0 11110000
49 31 00110001    113 71 01110001    177 b1 10110001    241 f1 11110001
50 32 00110010    114 72 01110010    178 b2 10110010    242 f2 11110010
51 33 00110011    115 73 01110011    179 b3 10110011    243 f3 11110011
52 34 00110100    116 74 01110100    180 b4 10110100    244 f4 11110100
53 35 00110101    117 75 01110101    181 b5 10110101    245 f5 11110101
54 36 00110110    118 76 01110110    182 b6 10110110    246 f6 11110110
55 37 00110111    119 77 01110111    183 b7 10110111    247 f7 11110111
56 38 00111000    120 78 01111000    184 b8 10111000    248 f8 11111000
57 39 00111001    121 79 01111001    185 b9 10111001    249 f9 11111001
58 3a 00111010    122 7a 01111010    186 ba 10111010    250 fa 11111010
59 3b 00111011    123 7b 01111011    187 bb 10111011    251 fb 11111011
60 3c 00111100    124 7c 01111100    188 bc 10111100    252 fc 11111100
61 3d 00111101    125 7d 01111101    189 bd 10111101    253 fd 11111101
62 3e 00111110    126 7e 01111110    190 be 10111110    254 fe 11111110
63 3f 00111111    127 7f 01111111    191 bf 10111111    255 ff 11111111
```

Index

& AND	107
& clears bits	111
&= AND equal	107
^ XOR	107
| OR	107
| OR equal	107
| sets bits	111
~ Bitwise complement	107
<<	107
>>	107
10 Types of People	104
1N4001	191
4N25 Optoisolator	183
Actuator	16
ALP	53
Amperes	78
analogWrite()	194
Arduino Jukebox Source Code	134
Arduino Learning Platform	53
Arduino Projects Kit	15
Arduino Volt Meter	82
Arithmetic Operators	102
ASCII	129, 205
ASCII_To_Integer Source Code	141
ATmega168/328 pins	98
ATmega328	26
atoi()	140
attachInterrupt()	179
AVM_Test Source Code	84
AVR Ports	99
AVRFreaks	20
Banzi, Massimo	14, 19
Bill of Materials	16
Binary	105, 106, 108
Bits	104, 107
Bitwise AND	107
Bitwise complement	107
Bitwise Operators	107
Bitwise OR	107
Bitwise vs Boolean	112
Blink Source Code	28
Blocks	45
Boolean	112
Boolean Operators	102
Bootloader	26, 30
Breadboards	47
Bytes	105
C Programming Language	38, 140
C Standard Libraries	140
C# Express	82
Cadmuim Sulfide (CdS	150
CdS Light Sensor	150
Centigrade	154
char	129
Circuits	80
Command_Demo Source Code	144
Comments	36, 43
Communication	129
Comparison Operators	102
Compound Operators	102
Constants	36
Conventional Current	81
Converting Centigrade to Fahrenheit	154
Cummutator	189
Cute L'il Bunny	162
Cylon Eyes	116
Cylon Eyes 1 Source Code	62
Cylon Eyes 2 Source Code	100
Cylon Eyes the Arduino Way	62

Index

CylonOptometry Source Code .. 119
Data Type 105
Data, Storing 152
DC Motor 189
Debounce 45, 103
Definitions 16
digitalRead() 45
Diode 191
DIP switch 117
DIP Switch 60
DIP to LED Source Code 61
DIPLED_With_Ports Source Code 99
Duemilanove 26
EAGLE 50
Edge Detection 178
Edge_Detect_Interrupt Source Code
 180
Electric Current 78
Electric Potential Difference 77
Electric Resistance 79
Electricity is dangerous (well, duh!)
 75
Encapsulation 44
Encoder Wheel 198
Exploratorium 14
Expressions 45
External Interrupts 178
Fade 72
Fahrenheit 154
FAQ 21
Flow Control 45
Foam Core Board 55
Franklin, Benjamin 81
FT232R 19
Functions 43
Greater Than 103
Happy Birthday Source Code 70

Help 30
Hexadecimal 106
I/O ports 97
IDE 18
IMHO 21
Increment 103
Infrared Object Detection 162
Installation 23
Interrupts 178
IR Reflective Object Sensor 165
Isolation 182
Language Reference 31
Launch_Control Source Code 114
LED 106
LED Dimmer 93
LEDs 105, 106
Left shift 107
Light Switch 39
Light_Switch - source code 41
LM35 Temperature Senso 155
LM35_Temperature Source Code 159
loop() 37, 44
Macros 116
Making IR visible 165
Masks 116
Microcontroller I/O Ports 97
Motor 189
Music 66
NOT 107
Number_Commander Source Code
 132
Nuts&Volts 19
Ohm's Law 79, 84
Ohms 79
Operators 102, 107
Optical Isolation 182
Parts List 16

Index

Physical Computing 17
Piezo Element 66
pinMode() 31
Potentiometer 88
Prerequisites 20
Processing 18
Program structure 36
Program writing 35
Pulse Width Modulation 194
PWM 72
PWM Motor Control 194
PWM_Test Source Code 185
QRD1114 165
Quick Start Guide 23
ReadPot Source Code 171
Receiving numbers 130
Right shift 107
Rotor 189
Schematic 50
Sending numbers 130
Sensor 16
Sensors 149
Serial Port 129
setup() 36, 44
Shift Operators 115
Short Circuits 81
Simple_Motor_Speed_Control
 Source Code 199
Sketch 17
Sketch vs. Program 30
Smiley Micros
 Arduino Projects Kit 15
Smiley's Workshop 19
Sound 66
Source Code
 Arduino Jukebox 134
 ASCII_To_Integer 141
AVM_Test 84
Blink 28
Command_Demo 144
Cylon Eyes 1 62
Cylon Eyes 2 100
CylonOptometery 119
DIP to LED 61
DIPLED_With_Ports 99
Edge_Detect_Interrupt 180
HappyBirthday 70
Launch Control 114
Light_Switch 41
LM35_Temperature 159
Number_Commander 132
PWM_Test 185
ReadPot 171
Simple_Motor_Speed_Control 199
Tomato_Soup_Can_Counter ... 173
Statements 45
Stator 189
Storing Data 152
Switch 39
TeleType 129
Temperature 154
Tinkering 14
TIP115 196
Tomato Soup Can Counter .. 170
Tomato_Soup_Can_Counter Source
 Code 173
Transistor 196
Tunes 68
Upload 30
USART 130
USB Serial Port 19
Using Internet Forums 33
Variables 36
Verify 17

211

Index

Verify vs. Compile 28
Volt Meter 82
Voltage 77
Voltage across resistance 84
Voltage Divider 84
Voltage Isolation 182
WinAVR 29
Wiring 18